Symmetry and Asymmetry in Quasicrystals or Amorphous Materials

Symmetry and Asymmetry in Quasicrystals or Amorphous Materials

Editor

Enrique Maciá Barber

MDPI • Basel • Beijing • Wuhan • Barcelona • Belgrade • Manchester • Tokyo • Cluj • Tianjin

Editor
Enrique Maciá Barber
Universidad Complutense de Madrid
Spain

Editorial Office
MDPI
St. Alban-Anlage 66
4052 Basel, Switzerland

This is a reprint of articles from the Special Issue published online in the open access journal *Symmetry* (ISSN 2073-8994) (available at: https://www.mdpi.com/journal/symmetry/special_issues/Symmetry_Asymmetry_Quasicrystals_Amorphous_Materials).

For citation purposes, cite each article independently as indicated on the article page online and as indicated below:

LastName, A.A.; LastName, B.B.; LastName, C.C. Article Title. *Journal Name* **Year**, *Article Number*, Page Range.

ISBN 978-3-03943-056-7 (Hbk)
ISBN 978-3-03943-057-4 (PDF)

© 2020 by the authors. Articles in this book are Open Access and distributed under the Creative Commons Attribution (CC BY) license, which allows users to download, copy and build upon published articles, as long as the author and publisher are properly credited, which ensures maximum dissemination and a wider impact of our publications.

The book as a whole is distributed by MDPI under the terms and conditions of the Creative Commons license CC BY-NC-ND.

Contents

About the Editor vii

Enrique Maciá Barber
Symmetry and Asymmetry in Quasicrystals or Amorphous Materials
Reprinted from: *Symmetry* **2020**, *12*, 1326, doi:10.3390/sym12081326 1

Jianhang Yue, Yun Feng, Hao Wu, Guorong Zhou, Min Zuo, Jinfeng Leng and Xinying Teng
The Study of A New Symmetrical Rod Phase in Mg-Zn-Gd Alloys
Reprinted from: *Symmetry* **2019**, *11*, 988, doi:10.3390/sym11080988 5

Vicenta Sánchez and Chumin Wang
Real Space Theory for Electron and Phonon Transport in Aperiodic Lattices via Renormalization
Reprinted from: *Symmetry* **2020**, *12*, 430, doi:10.3390/sym12030430 15

Edmundo Lazo
Localization Properties of Non-Periodic Electrical Transmission Lines
Reprinted from: *Symmetry* **2019**, *11*, 1257, doi:10.3390/sym11101257 37

K. Lambropoulos and C. Simserides
Tight-Binding Modeling of Nucleic Acid Sequences: Interplay between Various Types of Order or Disorder and Charge Transport
Reprinted from: *Symmetry* **2019**, *11*, 968, doi:10.3390/sym11080968 73

Hao Jing, Jie He, Ru-Wen Peng and Mu Wang
Aperiodic-Order-Induced Multimode Effects and Their Applications in Optoelectronic Devices
Reprinted from: *Symmetry* **2019**, *11*, 1120, doi:10.3390/sym11091120 99

About the Editor

Enrique Maciá Barber earned his BA degree in Physics from the Universidad Complutense de Madrid in 1987 and earned the excellence award for his PhD from the same university in 1995. Dr. Maciá is currently Full Professor of condensed matter physics at the Universidad Complutense de Madrid and has served as advisor in several national institutions, like the Real Sociedad Española de Física (Spanish Royal Society of Physics) and the Real Academia de Ciencias Exactas, Físicas y Naturales (Spanish Royal Academy of Sciences). In 2010, Prof. Maciá received the RSEF-BBVA Foundation Excellence Physics Teaching Award. His research interests cover different topics including electronic, thermal, and optical transport properties in aperiodic multilayers and DNA-based polymers; the nature of chemical bonding in quasicrystalline alloys; and the primary source of phosphorus compounds in chemical evolution. He is also author of several monographs and the books Thermoelectric Materials: Advances and Applications http://www.crcpress.com/product/isbn/9789814463522; Aperiodic Structures in Condensed Matter: Fundamentals and Applications http://www.crcpress.com/product/isbn/9781420068276; and The Chemical Evolution of Phosphorus. An Interdisciplinary Approach to Astrobiology http://www.appleacademicpress.com/the-chemical-evolution-of-phosphorus-an-interdisciplinary-approach-to-astrobiology/9781771888042. He is currently conducting a long-term research program on the role of aperiodic order in nature and leading several research projects on the thermoelectric properties of quasicrystals.

Editorial

Symmetry and Asymmetry in Quasicrystals or Amorphous Materials

Enrique Maciá Barber

Departamento Física de Materiales, Facultad CC. Físicas, Universidad Complutense de Madrid, 28040 Madrid, Spain; emaciaba@fis.ucm.es

Received: 3 August 2020; Accepted: 4 August 2020; Published: 9 August 2020

Abstract: Quasicrystals (QCs) are long-range ordered materials with a symmetry incompatible with translation invariance. Accordingly, QCs exhibit high-quality diffraction patterns containing a collection of discrete Bragg reflections. Notwithstanding this, it is still common to read in the recent literature that these materials occupy an intermediate position between amorphous materials and periodic crystals. This misleading terminology can be understood as probably arising from the use of models and notions borrowed from the amorphous solid's conceptual framework (such us tunneling states, weak interference effects, variable range hopping, or spin glass) in order to explain certain physical properties observed in QCs. On the other hand, the absence of a general, full-fledged theory of quasiperiodic systems certainly makes it difficult to clearly distinguish the features related to short-range order atomic arrangements from those stemming from long-range order correlations.

The Special Issue on "Symmetry and Asymmetry in Quasicrystals or Amorphous Materials" aims to discuss both experimental and fundamental aspects related to the relationship between the underlying structural order and the resulting physical properties of QCs and their related approximant phases [1], focusing on the analogies and differences between these properties and those reported for amorphous materials of similar composition.

It is currently agreed that the presence of non-crystallographic axes is not a necessary condition for a solid to be regarded as a QC, and that the key feature to this end is just to exhibit scale invariance symmetry [2,3]. Indeed, several examples of QCs exhibiting 2-, 3-, and 4-fold symmetry axes along with scale invariance symmetry characterized by irrational scale factors have been reported, being referred to as "cubic QCs" [4]. These findings support the view that QC definition should not include the requirement that they must display a classically forbidden axis of symmetry, as it is stated in the Online Dictionary of Crystallography of the International Union of Crystallography, where one reads: *"Often, quasicrystals have crystallographically 'forbidden' symmetries [...]. However, the presence of such a forbidden symmetry is not required for a quasicrystal"* [5].

The contribution by Jianhang Yue, Yun Feng, Hao Wu, Guorong Zhou, Min Zuo, Jinfeng Leng and Xinying Teng, entitled *The Study of A New Symmetrical Rod Phase in Mg-Zn-Gd Alloys*, nicely fits within this scenario. In this paper, the morphology and properties of Mg-Zn-Gd alloys prepared by a conventional casting method are studied by systematically varying their Mg and Gd content. In this way, a rod phase with atomic composition $Mg_{66}Zn_{30}Gd_4$ is reported to exhibit diffraction spots patterns indicating this phase belongs to a new kind of complex metallic alloy phase whose composition is close to that of $Mg_{60}Zn_{30}Gd_{10}$ QCs. Upon annealing, this rod phase evolved gradually over time from a lamellar eutectic structure, the melting temperature of the rod phase being 453°C. Quite interestingly from the viewpoint of possible applications, microhardness tests showed that its tribological properties are better than those corresponding to QCs of similar composition.

During the last decades we have realized that the electronic structure and vibrational spectrum of many quasiperiodic systems can be understood in terms of resonance effects involving a relatively small number of atomic clusters of progressively increasing size. In earlier works this scenario was discussed in terms of real-space based renormalization group approaches describing the mathematically simpler, but chemically unrealistic, diagonal (different types of atoms connected by the same type of bond) or off-diagonal (the same type of atom but different types of bonds between them) models. Later on, an increasing number of works have been devoted to the mathematically more complex general case, in which both diagonal and off-diagonal terms are present in the system. In fact, since the properties of chemical bonds linking two different atoms generally depend on their chemical nature, any realistic treatment must explicitly consider that the aperiodic sequence of atoms along the chain naturally induces an aperiodic sequence of bonds in the considered solid. Indeed, a growing number of both experimental measurements and numerical simulation results highlight the important role of chemical bonding in the emergence of some specific physical properties of QCs. This sort of more realistic treatments are discussed in three contribution focusing on appealing representatives of different kinds of aperiodic systems. In the contribution by Vicenta Sánchez and Chumin Wang, entitled *Real Space Theory for Electron and Phonon Transport in Aperiodic Lattices via Renormalization*, the unavoidable presence of structural defects are inherent in both periodic solids and QCs at a finite temperature is addressed by means of a real-space renormalization method, which uses an iterative procedure with a small number of effective sites in each step, and exponentially lessens the degrees of freedom, but keeps their participation in the final results. In this way, different aperiodic atomic arrangements with hierarchical symmetry are investigated, along with their consequences in measurable physical properties, such as electrical and thermal conductivities, in line with previous works of this research group [6,7].

In the contribution by Edmundo Lazo, entitled *Localization Properties of Non-Periodic Electrical Transmission Lines*, the properties of localization of the I(ω) electric current function in non-periodic electrical transmission lines are studied in detail. The electric components have been distributed in several forms: (a) Aperiodic, including self-similar sequences (Fibonacci and Thue–Morse), (b) incommensurate sequences (Aubry–André and Soukoulis–Economou), (c) long-range correlated sequences, and (d) uncorrelated random sequences. The localization properties of the transmission lines were measured by means of typical diagnostic tools of extended use quantum mechanics like normalized localization length, transmission coefficient, or average overlap amplitude, thereby exploiting the analogies between classic electric transmission lines and one-dimensional tight-binding quantum models.

In recent years thrust has been given to understand the spectral properties and the complexity of low dimensional systems, ranging from typical condensed matter systems in the mesoscopic or nano scales of length to biological systems. A common approach in most of these studies has been a kind of unified description involving quantum lattice models to explore diverse physical systems like DNA molecules, graphene nano-ribbons, fractals, hierarchical lattices, QCs or tribological systems. Long range topological order and chemical diversity in one dimensional, or quasi-one dimensional models and hierarchical lattices have shown to result into unusual spectral features like coexistence of extended (conducting) and localized electronic states, or even metal-insulator transitions in quasi-one dimensional ladder networks. The latter has also been successfully utilized to bring out the essential electronic structure and transport properties of DNA with periodic or aperiodic ordering of its constituents. Accordingly, some properties that are understood or claimed as specific properties of the systems considered may turn out to be quite general consequences of the adopted model instead. This subtle point requires a thorough inspection of various quantum lattice models addressing different physical systems.

This issue is fully addressed by Konstantinos Lambropoulos and Constantinos Simserides in the contribution entitled *Tight-Binding Modeling of Nucleic Acid Sequences: Interplay between Various Types of Order or Disorder and Charge Transport*. In their review tight-binding modeling of nucleic acid sequences like DNA and RNA is addressed by considering how various types of order (periodic, quasiperiodic, fractal) or disorder (diagonal, non-diagonal, random, methylation effects) affect charge transport. Several DNA models widely considered in the recent literature (wire, ladder, extended ladder, fishbone (wire), fishbone ladder) are considered within the framework of renormalization techniques. In doing so, the energy structure of nucleic acid wires, the coupling to the leads, the transmission coefficients and the current–voltage curves are numerically derived and the obtained results are discussed in order to examine the potentiality to utilize the charge transport characteristics of nucleic acids as a tool to probe several diseases.

Certainly, the very notion of photonic crystal can be extended to describe the properties of quasiperiodic photonic structures as well. To this end, one simply considers that the optical properties of the medium are given by a QP refraction index function, instead of a periodic one. The resulting structure can then be properly referred to as a photonic quasicrystal (PQC). Long-range quasiperiodic order, by its own, endows PQCs with certain characteristic properties which are not exhibited by their periodic counterparts. This feature stems from the richer structural complexity of aperiodic sequences, which arises from the presence of quasiperiodic and self-similar order related fingerprints, and naturally leads to the presence of a lot of resonant frequencies due to multiple interference effects throughout the structure. For instance, due to their highly fragmented frequency spectrum, aperiodic multilayers offer more full transmission peaks (alternatively, absorption dips) than periodic ones in a given frequency range for a given system length, and the inflation symmetry gives rise to a denser Fourier spectrum structure in reciprocal space. Thus, aperiodic photonic micro/nanostructures usually support optical multimodes. In the contribution by Hao Jing, Jie He, Ru-Wen Peng, and Mu Wang, entitled *Aperiodic-Order-Induced Multimode Effects and Their Applications in Optoelectronic Devices*, the authors review some developments of aperiodic-order-induced multimode effects and their applications in optoelectronic devices. It is shown that self-similarity or mirror symmetry in aperiodic micro/nanostructures can lead to optical or plasmonic multimodes in a series of one-dimensional/two-dimensional photonic or plasmonic systems. These multimode effects have been employed to achieve optical filters for the wavelength division multiplex, open cavities for light–matter strong coupling, multiband waveguides for trapping "rainbow", high-efficiency plasmonic solar cells, and transmission-enhanced plasmonic arrays. By all indications, these investigations will be beneficial to the development of integrated photonic and plasmonic devices for optical communication, energy harvesting, nanoantennas, and photonic chips.

Conflicts of Interest: The author declares no conflict of interest.

References

1. Maciá-Barber, E. Chemical bonding and physical properties in quasicrystals and their related approximants: Known facts and current perspectives. *Appl. Sci.* **2019**, *9*, 2132. [CrossRef]
2. Janssen, T.; Chapuis, G.; de Boissieu, M. *Aperiodic Crystals. From Modulated phases to Quasicrystals*; Oxford University Press: Oxford, UK, 2007.
3. Maciá-Barber, E. *Quasicrystals: Fundamentals and Applications*; Taylor & Francis, CRC Press: Boca Raton, FL, USA, 2020.
4. Elcoro, L.; Pérez-Mato, J.M. Cubic superspace symmetry and inflation rules in metastable MgAl alloy. *Eur. Phys. J. B* **1999**, *7*, 85–89. [CrossRef]
5. Available online: https://reference.iucr.org/dictionary/Quasicrystal (accessed on 1 August 2020).
6. Sánchez, F.; Sánchez, V.; Wang, C. Renormalization approach to the electronic localization and transport in macroscopic generalized Fibonacci lattices. *J. Non-Cryst. Solids* **2016**, *450*, 194. [CrossRef]
7. Wang, C.; Ramírez, C.; Sánchez, F.; Sánchez, V. Ballistic conduction in macroscopic non-periodic lattices. *Phys. Status Solidi B* **2015**, *252*, 1370–1381. [CrossRef]

© 2020 by the author. Licensee MDPI, Basel, Switzerland. This article is an open access article distributed under the terms and conditions of the Creative Commons Attribution (CC BY) license (http://creativecommons.org/licenses/by/4.0/).

Article

The Study of A New Symmetrical Rod Phase in Mg-Zn-Gd Alloys

Jianhang Yue, Yun Feng, Hao Wu, Guorong Zhou, Min Zuo, Jinfeng Leng and Xinying Teng *

School of Materials Science and Engineering, University of Jinan, No. 336, West Road of Nanxinzhuang, Jinan 250022, China
* Correspondence: mse_tengxy@ujn.edu.cn; Tel.: +86-137-9315-7586

Received: 8 July 2019; Accepted: 1 August 2019; Published: 2 August 2019

Abstract: Quasicrystal alloys have a wide application prospect because of excellent performances and characteristics; meanwhile, magnesium alloys are known as green engineering materials because of their high specific strength and light weight. Therefore, the study of Mg-Zn-Gd quasicrystal alloys is of great significance for the development of new materials. In this paper, $Mg_{(70-x)}Zn_{30}Gd_{x(x=3,4,5)}$ alloys were prepared by a conventional casting method and the morphologies and properties of these alloys were studied. There was a new symmetrical rod phase found in the $Mg_{66}Zn_{30}Gd_4$ alloy and the symmetrical rod phase was identified as a ternary phase by mapping scanning and energy dispersive spectroscopy (EDS) analysis. The Zn/Gd ratio of the symmetrical rod phase was found to be 4.8 and the TEM images obtained were different from the typical diffraction spots patterns of quasicrystalline, which means it is unlikely to be quasicrystalline. With different melt holding time, the symmetrical rod phase evolved gradually over time from a lamellar eutectic structure; differential scanning calorimetry (DSC), heat treatment, and microhardness tests showed that the melting temperature of the rod phase was 453 °C and that its thermal stability and microhardness are better than quasicrystalline. Hence, the symmetrical rod phase is a new kind of complex metallic alloy phase whose composition and properties are close to those of quasicrystals but is not quasicrystalline.

Keywords: Mg-Zn-Gd alloys; symmetrical rod phase; quasicrystal; morphologies and properties

1. Introduction

As the lightest green engineering structural material, magnesium alloy has many advantages, including high specific strength and stiffness, good shock and noise reduction performance, electromagnetic shielding, and easy processing and forming, etc. It has broad application prospects in transportation, aerospace, and military industries [1]. Shechtman et al. [2] first found the quasicrystalline phase in an Al-Mn quench alloy system in 1984. Unlike traditional crystals, quasicrystals have special symmetry of five or more times [3–6]. This structural particularity also makes them have high hardness and strength, as well as low friction coefficients, strong thermal stability, and corrosion resistance [7–9]. Therefore, introducing quasicrystals as a dispersion strengthening phase into magnesium alloys can theoretically compensate for the shortcomings of traditional magnesium alloys [10].

Quasicrystalline alloys have received more and more attention and recognition in recent decades [11,12]. In addition, studying the factors that affect the formation of quasicrystals for the synthesis and application of quasicrystals is significant. There are many factors affecting the formation of quasicrystals according to former studies, including cooling rate, composition, electronic structure, and melt treatment [13].

The quasicrystalline alloys of Mg-Zn-Re (where Re = rear earth element) have been extensively researched. It has been confirmed that the Zn/Y ratios of I-phase ($Mg_3Zn_6Y_1$) and W-phase ($Mg_3Zn_3Y_2$) are 6 and 1.5, respectively [14]. These ratios are in accordance with those for Mg-Zn-Gd alloys. In addition, W'-phase, H-phase, and other phases, which have different structures, exist in quasicrystalline alloys. Therefore, making a profound study of the structures and relationships between different phases will aid in obtaining a good understanding of the interlink among phases, which is of great significance to the study of the atomic structure of strengthened phases in high-performance magnesium alloys.

For Mg-Zn-Gd alloys, the icosahedral quasicrystal phase (I-phase) has been confirmed as being able to be made as an equilibrium phase during solidification or crystallization in a certain range of elemental components and holding time [15,16]. When the components or holding time change, the composition, structure, and symmetry of phases may change. In this paper, we prepared different $Mg_{(70-x)}Zn_{30}$-$Gd_{x(x=3,4,5)}$ (at. %) alloys and investigated the composition, structure, and symmetry of phases. This work focuses on the metastable phases in the Mg-Zn-Gd system; the formation mechanism of the symmetrical rod phase is also discussed.

2. Materials and Methods

The experimental ternary alloys with nominal component $Mg_{(70-x)}Zn_{30}Gd_{x(x=3,4,5)}$ (at. %) were prepared by melting high-purity Mg (99.98 wt %), Zn (99.96 wt %), and master alloy Mg-Gd (30.21 wt %) in an electric resistance furnace. Firstly, Mg and Mg-30.21 wt % Gd alloys were placed into a graphite-clay crucible. When the temperature reached 720 °C and the alloys previously added had melted, Zn was added into the molten metal. After all the alloys had become molten, the melt was kept at 720 °C for minutes. The melt was then poured into a 200 °C preheated steel mold and cooled slowly in the atmosphere. Protective gas composed of CO_2 and SF_6 was always in the process of smelting to prevent evaporation and oxidation of components during smelting.

X-ray diffraction (Bruker D8 Advance) employing Cu Kα radiation was used to determine the constitution of phases with a scanning rate of ~5° min^{-1}. The microstructure and morphology were characterized by SEM (FEI-QUANTA FEG250) equipped with EDS (X-MAX50) for analyzing the local chemical compositions of different phases and TEM (JEM-2100). The thermal stability and microhardness of phases was researched by DSC (HCT-1) and a Rockwell hardness tester. Sample preparations for TEM observations were made up of mechanical polishing and ion-beam thinning (GATAN-691).

3. Results

3.1. Microstructure and Composition

SEM images of Mg-Zn-Gd alloys with different content of Gd after holding for 40 min are displayed in Figure 1. The content of Gd in the Mg-Zn-Gd alloys seen in Figure 1a–c is 3 at. %, 4 at. %, and 5 at. %, respectively. It can be clearly seen that the phase morphologies of the alloys changed significantly. When the content of Gd is 3 at. % in the $Mg_{(70-x)}Zn_{30}$-Gd_x alloy (Figure 1a), three main phases are included in the alloy: a light grey pentapetaloid phase, a black punctate phase, and a dark grey matrix phase. According to the XRD patterns shown in Figure 2, these phases may be judged to be I-phase, α-Mg phase, Mg_7Zn_3 phase, respectively. This is consistent with the report in the literature by Gröbner et al. [17]. The TEM images of the light grey pentapetaloid phase (shown in Figure 5a,b) also appear to be a typical quasicrystal diffraction spot, which can also prove that the light grey pentapetaloid phase is an icosahedral quasicrystal (I-phase).

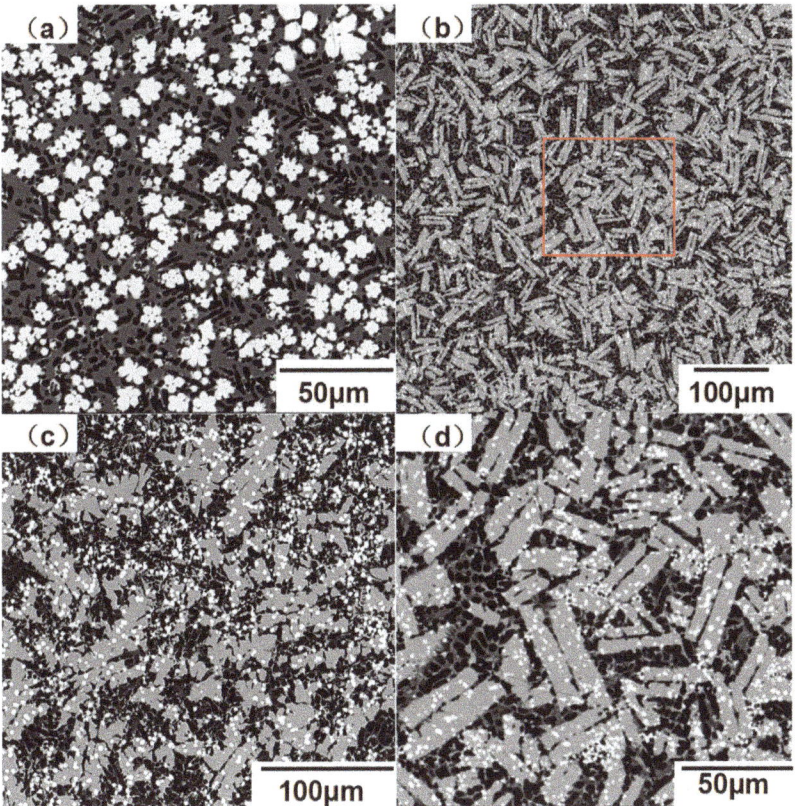

Figure 1. Mg-Zn-Gd alloys with different Gd content after holding for 40 min: (**a**) $Mg_{67}Zn_{30}Gd_3$; (**b**) $Mg_{66}Zn_{30}Gd_4$; (**c**) $Mg_{65}Zn_{30}Gd_5$; (**d**) locally enlarged image of (b).

Figure 1b is an SEM image of the $Mg_{66}Zn_{30}Gd_4$ alloy and Figure 1d is a local enlargement of Figure 1b. When the content of Gd reaches 4 at. %, the major phases except the black phase α-Mg and dark grey phase Mg_7Zn_3 are the light grey symmetrical rod phase and white punctate phase, which can be clearly seen from Figure 1d. The distribution of the symmetrical rod phase is disorderly and its length is 20 μm to 50 μm. The white punctate phase is dispersed in the alloy with a small volume. In the $Mg_{65}Zn_{30}Gd_5$ alloy (Figure 1c), the other three phases do not change a lot but the light grey phase appears irregularly shaped, which is quite different from the symmetrical rod phase.

As shown in the XRD patterns (Figure 2), when the content of Gd is 3 at. %, the peaks of the I-phase appear in the pattern of Figure 2a. However, the I-phase peaks are not detected in patterns of alloys with 4 at. % and 5 at. % Gd. On the contrary, in Figure 2b,c, phase Gd-Zn is found and there are obvious peaks in the places where 2θ is 23, 24.2, 36, 38, and 40.5, and these peaks are not the diffraction peaks of the other three phases. Thus, it can be inferred that these peaks may be the diffraction peaks of the symmetrical rod phase. Hence, deducing from Figures 1 and 2, the symmetrical rod phase is unlikely to be the icosahedral phase of quasicrystals.

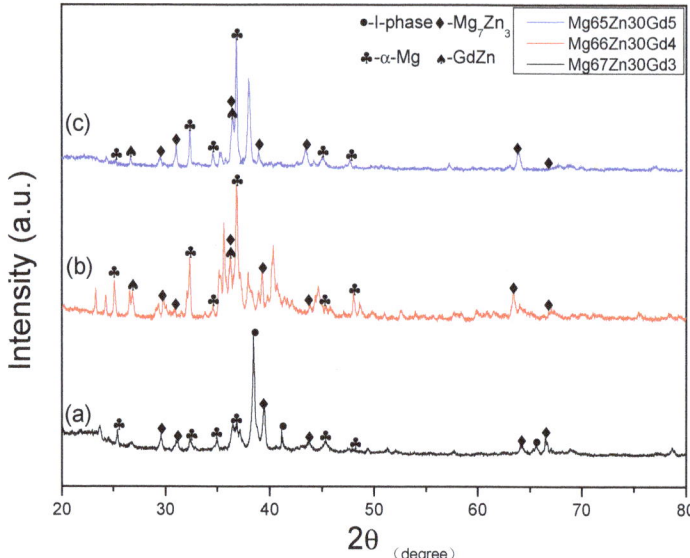

Figure 2. XRD patterns of different Mg-Zn-Gd alloys: (**a**) Mg$_{67}$Zn$_{30}$Gd$_3$; (**b**) Mg$_{66}$Zn$_{30}$Gd$_4$; (**c**) Mg$_{65}$Zn$_{30}$Gd$_5$.

Figure 3 is a mapping scanning analysis of the main morphologies of the Mg$_{66}$Zn$_{30}$Gd$_4$ alloy. It is obvious that the white punctate phase appears black in Mg and bright in the Zn and Gd graphs. This means that the white punctate phase contains almost no Mg element but embodies Zn and Gd, which is a kind of Gd-Zn alloy. By analyzing Figures 2 and 3, it can be approximately determined that the white punctate phase is the Gd-Zn phase. The white punctate phase Gd-Zn is mostly distributed over the symmetrical rod phase. Hence, the formation of this phase maybe due to the increase of Gd content, which means that the excess Gd element precipitates during the solidification of the alloy and reacts with Zn to form the Gd-Zn phase. In addition, the light grey symmetrical rod phase has bright colors in all three scanning images, meaning that the phase can be ascertained as a ternary phase containing Mg, Zn, and Gd elements.

To further research the phase composition, EDS analysis of the pentapetaloid I-phase and symmetrical rod phase was carried out. As shown in Figure 4, points 1 and 2 are the constituents of the pentapetaloid I-phase and the symmetrical rod phase, respectively. The atomic composition of the pentapetaloid I-phase is 65.82 at. % Mg, 29.22 at. % Zn, and 4.96 at. % Gd. Thus, it can be seen that the ratio of Zn/Gd in the pentapetaloid I-phase is 5.89, which is very close to the theoretical value Mg$_3$Zn$_6$Gd$_1$ [18] of the quasicrystalline phase in the Mg-Zn-Re system. The existence of the quintic rotational symmetry axis in Figure 5b and the EDS energy spectrum analysis further prove that the determination of the quintic petal phase as a quasicrystal phase is correct. The atomic composition of the symmetrical rod phase is 77.52 at. % Mg, 18.59 at. % Zn, and 3.89 at. % Gd, and the ratio of Zn/Gd is approximately 4.8. This is much less than the theoretical value of quasicrystalline.

Moreover, the TEM images of the symmetrical rod phase are shown in Figure 5c,d. It is obvious that the selected-area electron diffraction spots are very complex and different from the typical diffraction spots pattern (Figure 5b) of quasicrystalline; it may include various phases judging by the signed rectangles of different colors, but it does not include the quasicrystal phase. Jiang et al. [19] have reported that the ratio of Zn/Gd in the W-phase, which is common in Mg-Zn-Re alloys and is similar to the quasicrystalline phase, is about 1.5. As per the previous analysis, the Zn/Gd ratio of the symmetrical rod phase is 4.8, which is not only far from 1.5 (the ratio of W-phase), but also different from 6 (the ratio of I-phase). Therefore, by analyzing the SEM, XRD, EDS, and TEM graphics of the symmetrical rod phase, the composition of this phase is similar to that of quasicrystals, but its structure does not seem to have the characteristics of typical quasicrystals.

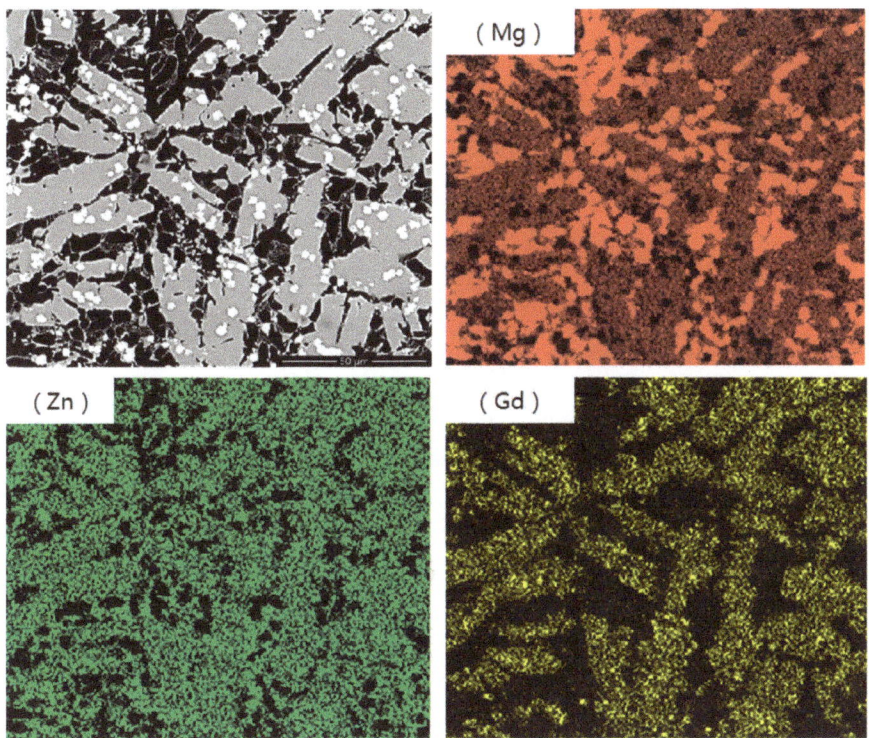

Figure 3. Mapping scanning analysis of the $Mg_{66}Zn_{30}Gd_4$ alloy.

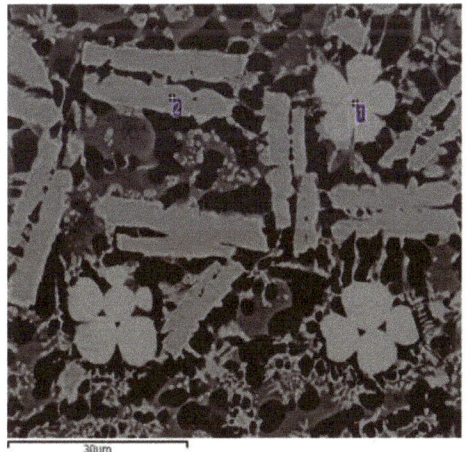

P_1

Element	At%
Mg	65.82
Zn	29.22
Gd	4.96

P_2

Element	At%
Mg	77.52
Zn	18.59
Gd	3.89

Figure 4. EDS analysis of different phases: (1) pentapetaloid phase; (2) symmetrical rod phase.

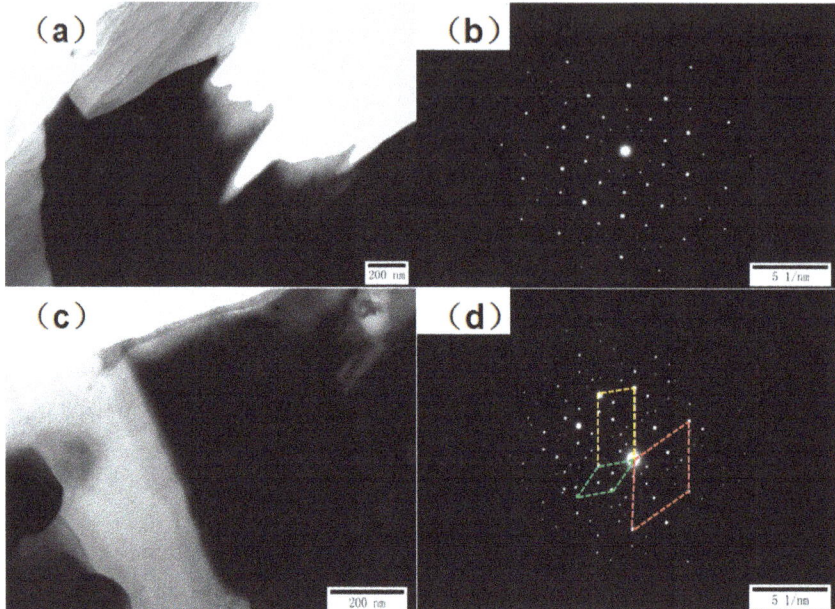

Figure 5. TEM images of different phases: (**a**,**b**) pentapetaloid phase; (**c**,**d**) symmetrical rod phase.

3.2. Morphological Evolution

In order to investigate the formation and stability of the symmetrical rod phase, samples of the $Mg_{66}Zn_{30}Gd_4$ alloy with different holding time at 720 °C were selected for research. As shown in Figure 6a, when the holding time is 5 min, most of the phases in the alloy are strip-like lamellar eutectic phases with a longitudinal midline running through the structure; meanwhile, there are some black α-Mg phases and white dotted Gd-Zn phases, which show that the GdZn phase is easy to form in the solidification of the alloy. After being held for 10 min (Figure 6b), the strip-like lamellar eutectic phases were less than before, and the white dotted phases did not increase significantly; however, the symmetrical rod phases had preliminarily formed. In the blue area of Figure 6c, the symmetrical rod phases had basically formed, and in the red area, it was half rod phase and half lamellar eutectic structure. Hence, it can be inferred that the symmetrical rod phase evolved gradually over time from the lamellar eutectic structure with the longitudinal midline as the demarcation line. When the holding time was 15 min, the evolution of the symmetrical rod phases had finished, and there was no lamellar eutectic structure which remained (Figure 6d). In Figure 6e, the size of the symmetrical rod phase increased to 40–70 μm but the morphologies essentially remained unchanged, which reflected the good stability of the symmetrical rod phase.

From Figure 7, it can be seen that the XRD diffraction patterns of the alloys after holding for 5, 10, and 15 min are basically the same and they all contain three identical known phases: Mg_7Zn_3, Gd-Zn and α-Mg. At the same time, peaks appear at positions 23, 24.2, 36, and 40.5 2θ, which is consistent with Figure 2b. Based on the analysis of Figures 6 and 7, it can be inferred that the symmetrical rod phase appeared early in the alloy, and gradually evolved from the lamellar network structure to the symmetrical rod phase structure.

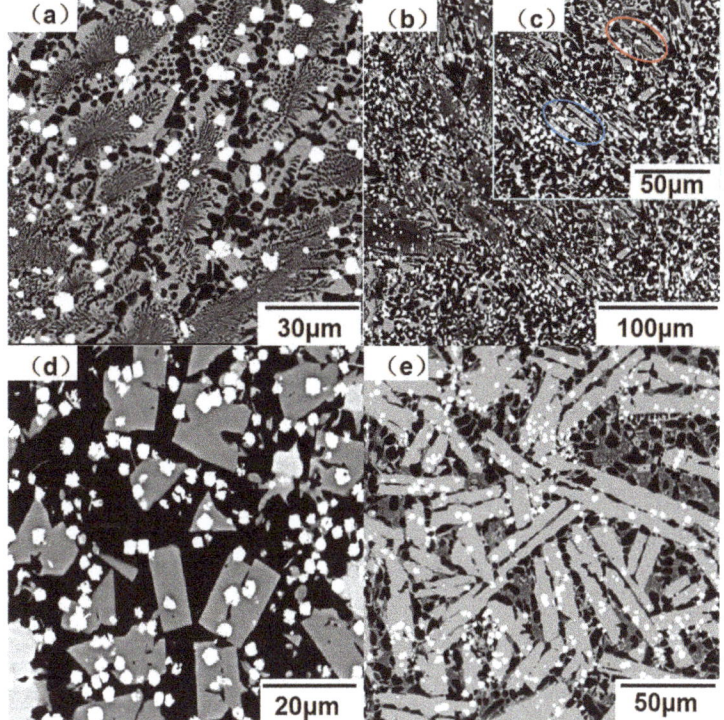

Figure 6. SEM images of the $Mg_{66}Zn_{30}Gd_4$ alloy after different holding times at 720 °C: (**a**) 5 min; (**b**) 10 min; (**c**) magnification of (b); (**d**) 15 min; (**e**) 80 min.

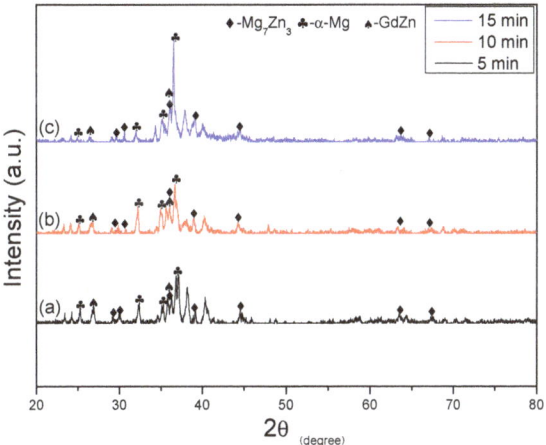

Figure 7. XRD patterns of the $Mg_{66}Zn_{30}Gd_4$ alloy after different holding times at 720 °C: (**a**) 5 min; (**b**) 10 min; (**c**) 15 min.

3.3. Thermal Stability

In addition, thermodynamic and microhardness tests of the petaloid quasicrystal phase (I-phase) and the symmetrical rod phase were carried out. Figure 8a shows the DSC analysis curves of the

$Mg_{67}Zn_{30}Gd_3$ and $Mg_{66}Zn_{30}Gd_4$ alloys. It can be seen that when the temperature reaches 345 °C, both lines have endothermic peaks, which is due to the melting of Mg_7Zn_3. Another endothermic peak of the $Mg_{67}Zn_{30}Gd_3$ alloy appears at 417 °C, and according to the reports of Zhang et al. [16], this is the melting peak of the petal-like quasicrystal phase; meanwhile, the melting peak of the symmetrical rod phase in the $Mg_{66}Zn_{30}Gd_4$ alloy appears at 453 °C, which means that the symmetrical rod phase may have a better thermal stability than the petal-like quasicrystalline phase. In order to verify the above hypothesis, samples of the $Mg_{67}Zn_{30}Gd_3$ and $Mg_{66}Zn_{30}Gd_4$ alloys were heat-treated at 430 °C for study. The results shown in Figure 8b,c indicate that the morphology of the quasicrystalline phase changed dramatically as a result of melting and that the petal-like morphology became a lamellar network structure; however, the morphology of the symmetrical rod phase remained stable in the main and only a few lamellar eutectic structures occurred in the interior. The microhardness of the quasicrystalline and symmetrical rod phases before and after heat treatment were studied; the results show that the microhardness of the symmetrical rod phase did not decrease obviously but that that of the quasicrystalline phase decreased a lot. All the above analyses prove that the symmetrical rod phase has a better thermal stability than the petal-like quasicrystalline phase.

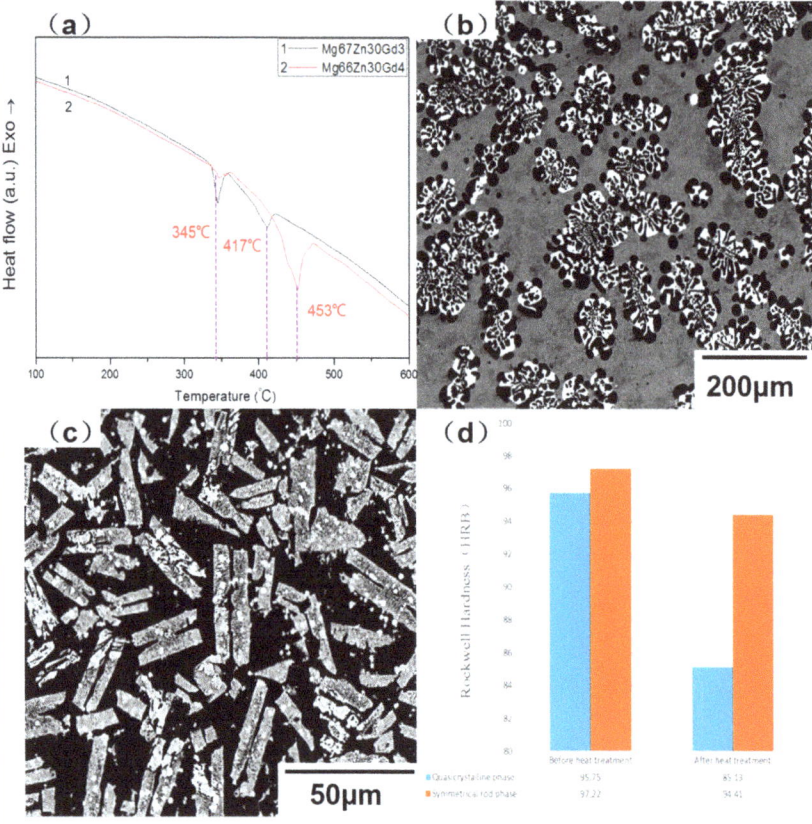

Figure 8. (a) Differential scanning calorimetry (DSC) analysis curves of the different Mg-Zn-Gd alloys: 1-$Mg_{67}Zn_{30}Gd_3$, 2-$Mg_{66}Zn_{30}Gd_4$; (b,c) SEM images of $Mg_{67}Zn_{30}Gd_3$ and $Mg_{66}Zn_{30}Gd_4$ after heat treatment at 430 °C, respectively; (d) microhardness of the quasicrystalline and symmetrical rod phases before and after heat treatment.

4. Discussion

In this study, the symmetrical rod phase was found in the $Mg_{66}Zn_{30}Gg_4$ alloy but not in the other two alloys considered, which indicates that the formation of the phase is related to the content of Gd, i.e., the ratio of Zn to Gd. The diffraction pattern of the symmetrical rod phase was found to be very complex, probably because it evolved from a lamellar eutectic structure, and its Zn/Gd ratio was observed to be close to the quasicrystal ratio, thus forming a more complex internal structure. Moreover, the increase of thermal stability and microhardness of the phase may also be due to the complex structure of the phase. Because the composition, Zn/Gd ratio, phase transition temperature, and microhardness of the symmetrical rod phase are similar to those of a quasicrystal phase, but its electron diffraction spot does not show the characteristics of a typical quasicrystal phase, we consider it to be a new kind of complex metallic alloy phase whose composition and properties are close to those of quasicrystals but is not quasicrystalline. The specific spatial structure and related parameters of the phase need to be further studied in the future.

5. Conclusions

In this work, the $Mg_{(70-x)}Zn_{30}Gd_{x(x=3,4,5)}$ alloys with different content of Gd were prepared and the morphologies and properties of these alloys were studied: (1) the main phases were found to be a pentapetaloid phase for the $Mg_{67}Zn_{30}Gd_3$, a symmetrical rod phase for the $Mg_{66}Zn_{30}Gd_4$ alloy, and an irregularly shaped phase for the $Mg_{65}Zn_{30}Gd_5$ alloy; (2) the symmetrical rod phase was found to consists of three elements (Mg, Zn, and Gd), its Zn/Gd ratio was obtained as 4.8 (which is close to but not in conformity with the requirements of quasicrystals) and its electron diffraction spots were complex and had no obvious quasicrystal phase characteristics; (3) this phase was found to evolve gradually over time from the lamellar eutectic structure and to be able to exist in alloy melt for a long time. The melting temperature of the symmetrical rod phase was 453 °C, and its thermal stability and microhardness were found to be better than those of quasicrystal phase; (4) the composition and phase transition temperature of the symmetrical rod phase were found to be close to those of a quasicrystal phase, but its electron diffraction pattern was seen to have no characteristics of a quasicrystal phase, so it may be a new kind of complex metal alloy phase, and needs further study.

Author Contributions: Conceptualization, J.Y.; data curation, Y.F.; formal analysis, J.Y.; methodology, Y.F.; software, H.W.; validation, G.Z., M.Z., and J.L.; writing—original draft, J.Y.; writing—review and editing, X.T.

Funding: This research was funded by the National Natural Science Foundation of China (grant nos. 51571102, 51701081, and 51871111), the Young-Aged Talents Lifting Project of the Shandong Association for Science and Technology (grant no. 301-1505001, recorded by the University of Jinan), and the Shandong Provincial Natural Science Foundation (grant no. ZR2017BEM001).

Acknowledgments: Thank you to my instructor, the teachers and students of the project group for their help, and also to the fundings.

Conflicts of Interest: The authors declare no conflict of interest.

References

1. Yuan, G.; Amiya, K.; Kato, H.; Inoue, A. Structure and mechanical properties of cast quasicrystal-reinforced Mg–Zn–Al–Y base alloys. *J. Mater. Res.* **2004**, *19*, 1531–1538. [CrossRef]
2. Shechtman, D.; Blech, I.; Gratias, D.; Cahn, J.W. Metallic phase with long-range orientational order and no translational symmetry. *Phys. Rev. Lett.* **1984**, *53*, 1951. [CrossRef]
3. Levine, D.; Steinhardt, P.J. Quasicrystals: A new class of ordered structures. *Phys. Rev. Lett.* **1984**, *53*, 2477. [CrossRef]
4. Fischer, S.; Exner, A.; Zielske, K.; Perlich, J.; Deloudi, S.; Steurer, W.; Lindner, P.; Förster, S. Colloidal quasicrystals with 12-fold and 18-fold diffraction symmetry. *Proc. Natl. Acad. Sci. USA* **2011**, *108*, 1810–1814. [CrossRef] [PubMed]

5. Gierer, M.; Hove, M.A.; Goldman, A.I.; Shen, Z.; Chang, S.L.; Jenks, C.J.; Zhang, C.M.; Thiel, P.A. Structural Analysis of the Fivefold Symmetric Surface of the A l 70 P d 21 M n 9 Quasicrystal by Low Energy Electron Diffraction. *Phys. Rev. Lett.* **1997**, *78*, 467. [CrossRef]
6. Zoorob, M.E.; Charlton, M.D.B.; Parker, G.J.; Baumberg, J.J.; Netti, M.C. Complete photonic bandgaps in 12-fold symmetric quasicrystals. *Nature* **2000**, *404*, 740–743. [CrossRef] [PubMed]
7. Gröbner, J.; Kozlov, A.; Fang, X.Y.; Geng, J.; Nie, J.F.; Schmid-Fetzer, R. Phase equilibria and transformations in ternary Mg-rich Mg–Y–Zn alloys. *Acta Mater.* **2012**, *60*, 5948–5962. [CrossRef]
8. Baake, M. *Quasicrystals: An Introduction to Structure, Physical Properties and Applications*; Suck, J.B., Schreiber, M., Häussler, P., Eds.; Springer: Berlin, Germany, 2002.
9. Vogel, M.; Kraft, O.; Dehm, G.; Arzt, E. Quasi-crystalline grain-boundary phase in the magnesium die-cast alloy ZA85. *Scr. Mater.* **2001**, *45*, 517–524. [CrossRef]
10. Zhang, J.; Jia, P.; Zhao, D.; Zhou, G.; Teng, X. Melt holding time as an important factor on the formation of quasicrystal phase in Mg67Zn30Gd3 alloy. *Phys. B Condens. Matter* **2018**, *533*, 28–32. [CrossRef]
11. Tanaka, R.; Ohhashia, S.; Fujitaae, N.; Demurabc, M.; Yamamotob, A.; Katod, A.; Tsai, A.P. Application of electron backscatter diffraction (EBSD) to quasicrystal-containing microstructures in the Mg-Cd-Yb system. *Acta Mater.* **2016**, *119*, 193–202. [CrossRef]
12. Jeon, S.Y.; Kwon, H.; Hur, K. Intrinsic photonic wave localization in a three-dimensional icosahedral quasicrystal. *Nat. Phys.* **2017**, *13*, 363. [CrossRef]
13. Huang, H.; Tian, Y.; Yuan, G.; Chen, C.; Ding, W.; Wang, Z. Formation mechanism of quasicrystals at the nanoscale during hot compression of Mg alloys. *Scr. Mater.* **2014**, *78*, 61–64. [CrossRef]
14. Huang, H.; Tian, Y.; Yuan, G.; Chen, C.; Ding, W.; Wang, Z. Precipitation of secondary phase in Mg-Zn-Gd alloy after room-temperature deformation and annealing. *J. Mater. Res. Technol.* **2018**, *7*, 135–141. [CrossRef]
15. Tian, Y.; Huang, H.; Yuan, G.; Chen, C. Nanoscale icosahedral quasicrystal phase precipitation mechanism during annealing for Mg–Zn–Gd-based alloys. *Mater. Lett.* **2014**, *130*, 236–239. [CrossRef]
16. Zhang, J.; Teng, X.; Xu, S.; Ge, X.; Leng, J. Temperature dependence of resistivity and crystallization behaviors of amorphous melt-spun ribbon of Mg66Zn30Gd4 alloy. *Mater. Lett.* **2017**, *189*, 17–20. [CrossRef]
17. Gröbner, J.; Kozlova, A.; Fang, X.Y.; Zhu, S.; Nie, J.F. Phase equilibria and transformations in ternary Mg–Gd–Zn alloys. *Acta Mater.* **2015**, *90*, 400–416. [CrossRef]
18. Sugiyama, K.; Yasuda, K.; Ohsuna, T.; Hiraga, K. The structures of hexagonal phases in Mg-Zn-Re (Re= Sm and Gd) alloys. *Z. Fur Krist.* **1998**, *213*, 537–543. [CrossRef]
19. Jiang, H.; Qiao, X.; Xu, C.; Kamado, S.; Wu, K.; Zheng, M. Influence of size and distribution of W-phase on strength and ductility of high strength Mg-5.1 Zn-3.2 Y-0.4 Zr-0.4 Ca alloy processed by indirect extrusion. *J. Mater. Sci. Technol.* **2018**, *34*, 277–283. [CrossRef]

© 2019 by the authors. Licensee MDPI, Basel, Switzerland. This article is an open access article distributed under the terms and conditions of the Creative Commons Attribution (CC BY) license (http://creativecommons.org/licenses/by/4.0/).

Review

Real Space Theory for Electron and Phonon Transport in Aperiodic Lattices via Renormalization

Vicenta Sánchez [1] and Chumin Wang [2],*

[1] Departamento de Física, Facultad de Ciencias, Universidad Nacional Autónoma de México, Ciudad de México 04510, Mexico; vicenta@unam.mx
[2] Instituto de Investigaciones en Materiales, Universidad Nacional Autónoma de México, Ciudad de México 04510, Mexico
* Correspondence: chumin@unam.mx; Tel.: +52-55-5622-4634

Received: 7 February 2020; Accepted: 3 March 2020; Published: 7 March 2020

Abstract: Structural defects are inherent in solids at a finite temperature, because they diminish free energies by growing entropy. The arrangement of these defects may display long-range orders, as occurring in quasicrystals, whose hidden structural symmetry could greatly modify the transport of excitations. Moreover, the presence of such defects breaks the translational symmetry and collapses the reciprocal lattice, which has been a standard technique in solid-state physics. An alternative to address such a structural disorder is the real space theory. Nonetheless, solving 10^{23} coupled Schrödinger equations requires unavailable yottabytes (YB) of memory just for recording the atomic positions. In contrast, the real-space renormalization method (RSRM) uses an iterative procedure with a small number of effective sites in each step, and exponentially lessens the degrees of freedom, but keeps their participation in the final results. In this article, we review aperiodic atomic arrangements with hierarchical symmetry investigated by means of RSRM, as well as their consequences in measurable physical properties, such as electrical and thermal conductivities.

Keywords: quasiperiodicity; localization; tight-binding model; Kubo formula; low-dimensional systems

1. Introduction

Nowadays, impurities and defects in solids play a central role in microelectronics and modern materials science, because they deeply alter the propagation and interference of electronic wave functions [1]. In general, structural disorder obstructs the transport of excitations. However, this obstruction to both electronic and phononic transport could become beneficial, such as for the thermoelectricity, whose figure of merit is a function of the ratio between electrical and thermal conductivities [2].

Since the formulation of quantum mechanics at the beginning of the last century, the study of crystalline solids is carried out through the reciprocal lattice and local imperfections are addressed as perturbations [3]. For extended random disorders, the coherent potential approximation (CPA) is used in their analysis [4]. The discovery of quasicrystals by D. Shechtman et al. in 1984 [5] has stimulated the development of new techniques to investigate the long range and hierarchically located impurities or defects. During many years, the quasiperiodic systems have been studied using approximants [6], whose artificial periodic boundary condition has deep effects on the entire band structure of a truly quasiperiodic lattice, as illustrated in Figure 1.

To address macroscopic aperiodic lattices, the traditional reciprocal space [7] approach becomes inappropriate or useless, as the aperiodicity collapses the first Brillouin zone. An alternative way could be the real-space renormalization method (RSRM) firstly proposed by Leo P. Kadanoff [8] for condensed matter physics in 1966, to realize a scaling analysis of magnetization in terms of spin blocks, which exponentially reduces the degrees of freedom, keeping only the lower energy states. In 1971, Kenneth G. Wilson [9,10] reformulated the RSRM to introduce the universality classes of scale-independent critical points in phase transitions and was awarded by the Nobel Prize in Physics for this work in 1982.

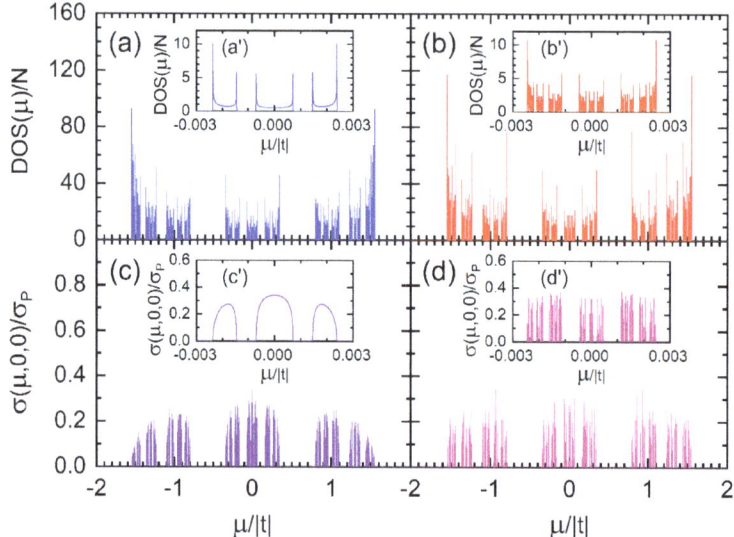

Figure 1. (color online) (**a**,**b**) Density of states (DOS) and (**c**,**d**) zero-temperature direct current (DC) conductivity (σ) versus the chemical potential (μ) for two bond-disordered Fibonacci chains (**b**,**d**) of $n = 57$ and (**a**,**c**) with a unit cell of $n = 15$. Insets (**a'**–**d'**) are the respective magnifications of (**a**–**d**) spectra.

In this article, we first introduce the tight-binding model and the Kubo–Greenwood formula [11] to describe the electronic transport in Fibonacci chains, as well as RSRM developed to reach macroscopic length. Other aperiodic chains, beyond the quasiperiodic ones, are further presented with a special emphasis on their electronic wave-function localization and the ballistic transport states. Studies on multidimensional aperiodic lattices are subsequently summarized, where the combination of RSRM with convolution theorem is shown. In Section 5, we discuss vibrational excitations or phonons in aperiodic lattices, as well as the thermoelectric transport in segmented heterostructures. Final remarks will be given in the conclusion section.

2. Fibonacci Chains

Let us first consider a single electron in a periodic lattice of atoms, which is usually addressed by means of Bloch's theorem [3]. This theorem establishes a general solution of the Schrödinger equation for a periodic potential, and then the electronic wave functions are commonly written as a linear combination of plane waves. Alternatively, such wave functions can also be expressed in terms of atomic orbitals, because they constitute another base for solutions of the Schrödinger equation. In fact, the orthonormalized orbitals of all atoms, known as Wannier functions, are the Fourier transformed Bloch functions [3].

For aperiodic lattices, the Wannier functions localized at each atom remain as a useful base. The single-band electronic Hamiltonian within the tight-binding formalism can be written as

$$H = \sum_j \varepsilon_j |j\rangle\langle j| + \sum_{\langle j,l \rangle} t_{j,l} |j\rangle\langle l| \qquad (1)$$

where ε_j is the self-energy of atom j with Wannier function $|j\rangle$ and $t_{j,l}$ is the hopping integral between the nearest-neighbor atoms j and l denoted by $\langle j,l\rangle$. The density of states (DOS) can be calculated using the single-electron Green's function (G) [12]:

$$DOS(E) = -\frac{1}{\pi}\lim_{\eta\to 0^+} \mathrm{Im} Tr[G(E+i\eta)] \qquad (2)$$

where η is the imaginary part of energy E and the Green's function is determined by the Dyson equation given by $(E-H)G = 1$.

Within the linear response theory, the electrical conductivity (σ) can be determined by means of the Kubo–Greenwood formula [11,12]:

$$\sigma_{xx}(\mu,\omega,T) = \frac{2e^2\hbar}{\Omega\pi m^2}\int_{-\infty}^{\infty} dE \frac{f(E)-f(E+\hbar\omega)}{\hbar\omega} Tr\left[p_x \mathrm{Im} G^+(E+\hbar\omega)p_x \mathrm{Im} G^+(E)\right] \qquad (3)$$

where Ω is the system volume, $p_x = (im/\hbar)[H,x] = (ima/\hbar)\sum_j\{t_{j,j+1}|j\rangle\langle j+1| - t_{j,j-1}|j\rangle\langle j-1|\}$ is the projection of the momentum operator along the applied electrical field with $x = \sum_j ja|j\rangle\langle j|$, $G^+(E) = G(E+i\eta)$ is the retarded Green's function, and $f(E) = \{1+\exp[(E-\mu)/k_BT]\}^{-1}$ is the Fermi–Dirac distribution with the chemical potential μ and temperature T. The electrical conductivity of direct current (DC) at zero temperature of a periodic linear chain ($t_{j,l} = t$) of N atoms with null self-energies is as follows [13]:

$$\sigma_P = \sigma(\mu,0,0) = \frac{(N-1)ae^2}{\pi\hbar} \qquad (4)$$

when the chemical potential is found in the allowed energy band, that is, $|\mu|\leq 2|t|$. It would be worth mentioning that the non-uniformity of atomic locations can be introduced through replacing the hopping integrals $t_{j,l}$ by $\tilde{t}_{j,l} = |x_j - x_l|t_{j,l}/a$ in the momentum operator expression.

The most studied quasiperiodic system is the Fibonacci chain, shown in Figure 2a, which can be built using two sorts of bonds (bond problem), two kinds of atoms (site problem), or a combination of both (mixing problem) [14]. For example, in the bond problem, the nature of atoms is assumed to be the same ($\varepsilon_j = 0$) and two bond strengths t_A and t_B are ordered following the Fibonacci sequence [15,16], whose atomic chain of generation n can be obtained using the concatenation of two previous generations, $F_n = F_{n-1} \oplus F_{n-2}$, with the initial conditions of $F_1 = A$ and $F_2 = AB$.

In Figure 1, we present (a,b) the density of states (DOS) and (c,d) the DC electrical conductivity at zero temperature (σ) as functions of the chemical potential (μ) for (a,c) a Fibonacci chain with bond disorder of $t_A = \frac{1}{2}(\sqrt{5}-1)t$ and $t_B = t$ made of a unit cell of generation $n = 15$ (987 bonds) repeated by $2^{29} = 536,870,912$ times, resulting a chain of $N = 529,891,590,145$ atoms connected to two leads built by repeating 2^{100} times the mentioned unit cell, and (b,d) a Fibonacci chain of generation $n = 57$ with $N = 591,286,729,880$ atoms having the same bond disorder strength as in (a,c). Both DOS and σ results were calculated by means of the renormalization method developed in [17] with grids of (a–d) 800,000 and (a'–d') 300,000 data. The imaginary parts of energy used in these figures are $\eta = 10^{-6}|t|$ for DOS and $\eta = 10^{-14}|t|$ for σ spectra.

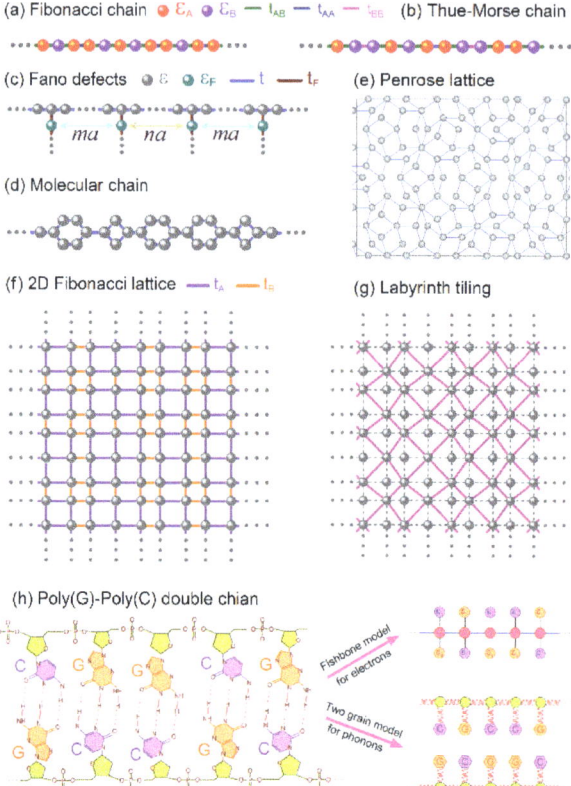

Figure 2. (color online) Schematic representations of one-dimensional (**a**) Fibonacci, (**b**) Thue-Morse, (**c**) branched, and (**d**) molecular chains, as well as two-dimensional (**e**) Penrose, (**f**) Fibonacci, (**g**) labyrinth, and (**h**) Poly(G)-Poly(C) lattices.

Observe the close similarity between *DOS* spectra shown in Figure 1a,b in contrast to the conductivity spectra of Figure 1c,d, as well as the remarkable differences between Figure 1a', b', where the continuum energy bands in Figure 1a' are originated from the periodic repetition of a unit cell. These differences can significantly modify the calculation of many physical quantities weighted by DOS spectra, such as the specific heat, optical absorption, and low-temperature DC and alternating current (AC) conductivities. Hence, the accurate determination of DOS and σ spectra constitutes a crucial starting point for the study of quasiperiodic systems.

The RSRM has been applied to quasiperiodic systems described by tight-binding Hamiltonian (1) since the discovery of quasicrystals. For example, from 1984 to 1987, M. Kohmoto and collaborators carried out renormalization group studies of Cantor-set electronic band spectra [18,19], the diffusion coefficient [20], localization properties [21,22], and the resistance power–law growth with sample length [23]. Q. Niu and F. Nori developed, in 1986, a decimation procedure to calculate energy spectra of Fibonacci chains based on the weak bond approximation [24], which was also applied to a scaling analysis of sub-band widths [25]. In 1988, H. E. Roman derived a RSRM to calculate on-site energies and hopping integrals of each generation [26], P. Villaseñor-González, F. Mejía-Lira, and J. L. Morán-López calculated the electronic density of states in off-diagonal Fibonacci chains [27], while C. Wang and R. A. Barrio obtained [28] the Raman spectrum measured in GaAs-AlAs quasiperiodic superlattices [29]. Moreover, the RSRM has also been used for the local electronic density of states [30],

Ising model [31,32], and alternating current (AC) conductivity [33] through the resistance network model of Miller and Abrahams [34].

In the 1990s, more attempts were registered to develop and use the renormalization technique. For example, J. C. López, G. G. Naumis, and J. L. Aragón determined [35] the electronic band structure of disordered Fibonacci chains following the renormalization procedure of Barrio and Wang [36]; while R. B. Capaz, B. Koiller, and S. L. A. de Queiroz studied the power–law localization behavior [37]; Y. Liu and W. Sritrakool found energy spectrum branching rules [38]; A. Chakrabarti et al. analyzed the nature of eigenstates [39]; and J. X. Zhong et al. calculated the local [40] and average [41] density of states. Besides, AC conductivity was examined [42,43] within the Miller and Abrahams approach. During the second half of the decade, F. Piéchon, M. Benakli, and A. Jagannathan established analytical scaling properties of energy spectra [44]; E. Maciá and F. Domínguez-Adame proved the existence of transparent states [45]; while A. Ghosh and S. N. Karmakar explored the second-neighbor hopping problem [46].

From the twenty-first century, the electronic transport in Fibonacci chains was deeply studied via renormalization. For instance, V. Sánchez et al. developed, in 2001, a sophisticated and exact RSRM for the Kubo–Greenwood formula (3) applied to the mixing Fibonacci problem [47], and then its AC conductivity spectra were carefully analyzed [48,49] beyond those obtained from approximants [50]. The renormalization technique was also used for the study of localization [51–53], electronic spectra of $GaAs/Ga_xAl_{1-x}As$ superlattices [54], and arrays of quantum dot [55], as well as for a unified transport theory of phonon [56], photon [57], and fermionic atom [58] based on the tight-binding model. On the other hand, by means of RSRM, the fine structure of energy spectra [59] and electronic transport in Hubbard Fibonacci chains [60,61] were investigated, and a new universality class was found in spin-one-half Heisenberg quasiperiodic chains [62].

3. Aperiodic Chains besides Fibonacci

Among aperiodic sequences, the generalized Fibonacci (GF) order was one of the most studied, which can be obtained by the substitutional rule:

$$A \to A^u B^v ? \text{and } B \to A \tag{5}$$

or using the substitution matrix (**M**):

$$\begin{pmatrix} A \\ B \end{pmatrix} \to \mathbf{M} \begin{pmatrix} A \\ B \end{pmatrix} = \begin{pmatrix} u & v \\ 1 & 0 \end{pmatrix} \begin{pmatrix} A \\ B \end{pmatrix} = \begin{pmatrix} \underbrace{AA \cdots}_{u} \underbrace{ABB \cdots B}_{v} \\ A \end{pmatrix} \tag{6}$$

where u and v are positive integer numbers. Matrix **M** has the following eigenvalues ($\lambda_\pm$):

$$\begin{vmatrix} u - \lambda & v \\ 1 & -\lambda \end{vmatrix} = 0 \Rightarrow \lambda^2 - u\lambda - v = 0 \Rightarrow \lambda_\pm = \frac{u \pm \sqrt{u^2 + 4v}}{2} \tag{7}$$

For $v = 1$, Equation (7) leads to $\lambda_+ > 1$ and $|\lambda_-| < 1$, which fulfill the Pisot conjecture [14,63]. Moreover, the determinant of **M**,

$$\det(\mathbf{M}) = \begin{vmatrix} u & v \\ 1 & 0 \end{vmatrix} = -v \tag{8}$$

is unimodular if $v = 1$. Hence, the corresponding sequences are called quasiperiodic and possess Bragg-peak diffraction spectra, because both the Pisot eigenvalue condition and the unit-determinant requirement of **M** are satisfied [64]. On the contrary, the GF sequences with $v \neq 1$ do not satisfy the unit-determinant requirement and thus they are not quasiperiodic. When $u = v = 1$, the sequence is called golden mean or the standard Fibonacci one, while the cases $u = 2$ and $u = 3$ are named silver and bronze means, respectively, when $v=1$, which are also known as the precious means. In addition, the metallic means stand for the sequences with $u=1$ and $v>1$ [65].

Since 1988, the electronic properties of GF chains have been investigated [66] and the RSRM was applied for calculating the average Green's function [67,68], local [69–73], and integrated [74,75] density of states, as well as for analyzing the localization of eigenstates [76]. Nonequilibrium phase transitions were analyzed by means of RSRM and Monte Carlo approaches [77]. Recently, the ballistic transport was found at the center of energy spectra in macroscopic GF chains with bond disorder every six generations when $v=1$ or all generations when u and v are both even numbers [78], whose wave-function localization and electrical conductivities (DC and AC) were investigated through a system length scaling analysis [79].

On the other hand, the Thue–Morse (TM) sequence constitutes another widely studied aperiodic order, whose nth generation chain, denoted by TM_n, can be constructed using the substitution rule $A \to AB$ and $B \to BA$, or the addition rule $TM_n = TM_{n-1} \oplus \overline{TM_{n-1}}$, where the symbol $\oplus$ stands the string concatenation and $\overline{TM_n}$ is the complement of TM_n, obtained by exchanging A and B in TM chains. The initial condition is $TM_0 = A$, and thus $TM_3 = ABBABAAB$ has 2^3 atoms, being the eight most left atoms in Figure 2b. The TM sequence accomplishes the Pisot conjecture, but it has a null substitution matrix determinant, $\det(\mathbf{M}) = 0$, as periodic lattices [80]. In consequence, it is not a quasiperiodic system, but exhibits an essentially discrete diffraction pattern, and then TM heterostructures can be regarded as an aperiodic crystal according to the definition of crystals given by the International Union of Crystallography [81]. The RSRM has been applied to the study of electronic properties in TM chains since 1990 [82], where the density of states [83], trace map problem [84,85], and localization [86,87], as well as excitonic states [88], were analyzed.

Another example of aperiodic sequence studied by RSRM was period doubling (PD), whose sequence can be generated by substitutions $A \to AB$ and $B \to AA$, or the addition rule $PD_n = PD_{n-1} \oplus PD_{n-2} \oplus PD_{n-2}$, where PD_n is the PD chain of generation n and the initial conditions are $PD_0 = A$ and $PD_1 = AB$. For example, $PD_2 = ABAA$ and $PD_3 = ABAAABAB$. The local [89] and global [90] electronic properties of pristine and random PD chains, as well as critical behavior of the Gaussian model [91], were studied via RSRM. Moreover, three-component Fibonacci chains, defined by the inflation rules $A \to B$, $B \to C$, and $C \to CA$, were addressed by the RSRM, where branching rules of their electronic energy spectra were analytically obtained [92] and compared with the numerical local density of states [93]. A summary on the nature of electronic wave functions in one-dimensional (1D) aperiodic lattices can be found in [94].

4. Multidimensional Aperiodic Lattices

Beyond one-dimensional systems, let us first consider a linear chain with branches of atoms, known as Fano-Anderson defects [95], which is illustrated in Figure 2c and has an average coordination number of larger than two, but without loops. The appearance of such branches may significantly modify the transport of excitations along the linear chain owing to the wave interference. In fact, quasiperiodically placed branches could inhibit the transport of long-wavelength excitations, which are usually unaltered by local impurities or defects [96]. Electronic transport in a quantum wire with an attached quantum-dot array was studied by P. A. Orellana et al. in 2003 [97], while engineering Fano resonances in discrete arrays were proposed by A. E. Miroshnichenko and Y. S. Kivshar in 2005 [98]. During the next decade, more detailed studies using RSRM were carried out for the transmission coefficient [99–105], Landauer resistance [106], Lyapunov exponent [100], local DOS [101–103], and Kubo conductivity [107]. Moreover, the ballistic AC conductivity of periodic lattices has been surpassed through quasiperiodicity [108] or Fano resonances [109].

Linear chains built by ring molecules, illustrated in Figure 2d, constitute another example of systems with an effective dimensionality bigger than one, whose atomic loops produce a rich quantum interference of the conducting wavefunction. This interference enables high-performance molecular switching with large on/off ratios essential for the next generation of molecular electronics [110,111], where the RSRM has been used for the study of Fibonacci arrays of Aharonov–Bohm rings [112], metal–insulator transition in the quasiperiodic Aubry model [113], electronic transmission in bent quantum wires [114], and in ladders with a single side-attached impurity [115]. Recently, the electronic density of states, localization, transmittance, and persistent current in molecular chains and ladders have been widely investigated via RSRM [116–122], while the spin-selective electronic transport was also analyzed [123,124]. A review of these studies is presented in [125].

Self-assembled deoxyribonucleic acid (DNA) molecular wires, built by cytosine-guanine (CG) or adenine-thymine (AT) stacked pairs attached to the double-helix structure through sugar-phosphate backbones, may behave as a low-dimensional conductor, semiconductors, or insulators, depending on the system length and base-pair sequences [126,127]. Ab-initio [128,129] and semi-empirical [130,131] studies of DNA molecules were carried out and, among them, the latter has the advantage of being simple and suitable for the analysis of electronic transport in aperiodic double chains with macroscopic length. The DNA molecules can be modelled as a double-strand ladder of coarse grains, which has been transformed into a single string of base pairs with dangling backbones, known as the fishbone model, and in turn, it was reduced to a single chain after a two-step renormalization at each base pair [132]. This chain has been used for the study of electronic transport in Fibonacci [133,134] and asymmetric [135] DNA molecules, helical structures [136–138], thermoelectric devices [139], diluted random base-pair segments [140], and Hubbard systems [141]. An additional renormalization process can be carried out along organic molecular wires to calculate the density of states [142,143], Lyapunov coefficient [144,145], transmittance [142–145], and magnetoconductance [146]. In fact, the double-strand ladder model is still used for the analysis of charge transport in quasiperiodic Poly (CG) systems [147] and a comparison between ladder and fishbone models was also performed [148]. Moreover, a possible test of the Efimov states in three-strand DNA systems was proposed [149,150]. Several review articles about DNA-based nanostructures have recently been published [151,152].

A two-dimensional (2D) square Fibonacci lattice can be constructed by superimposing two 1D Fibonacci chains along the x and y axis, as shown in Figure 2f, whose Hamiltonian could be defined as $H^{2D} = H_x^{1D} \otimes I_y^{1D} + I_x^{1D} \otimes H_y^{1D}$ with H_ν^{1D} (I_ν^{1D}) the 1D Hamiltonian (identity matrix) along the $\nu = x$ or y axis. Hence, for the bond problem, this construction procedure is straightforward [153], while three kinds of sites are generated in the site or mixing problems [154]. A special case of Fibonacci superlattices is obtained when one of these chains is quasiperiodic and another is periodic, in which the 2D problem can be addressed by applying the reciprocal space technique along the periodic direction and the renormalization method along the quasiperiodic one [155]. For the three-dimensional (3D) case, a Fibonacci superlattice is generally obtained from a 2D periodic lattice and a 1D quasiperiodic one, as occurring in the quasiperiodic GaAs-AlAs heterostructure constructed by R. Merlin et al. [156], whose vibrational spectrum was calculated by a combined method of real and reciprocal spaces [28]. In the last three decades, the splitting rules of electronic energy spectra [157–159], density of states [160,161], and DC [162–164] and AC [165,166] electrical conductance in 2D Fibonacci lattices have been extensively studied.

For 3D aperiodic systems with a small cross section, that is, non-periodic nanowires, the electrical conductance [167,168] and impurity effects [169,170] were investigated by means of the renormalization plus convolution technique of [17], whose computational efficiency is shown in Figure 3 and compared to the direct calculation through the matrix inversion process. The computing times shown in Figure 3 correspond to the calculations of zero-temperature DC conductivity given by Equation (3) at $\mu = 0$ for a quasiperiodic nanowire with a cross section of 5×5 atoms, where the Fortran's quadruple precision and a Supermicro workstation with two central processing unit (CPU) processors of Intel Xeon 4108 and 64 GB of DDR4-2666 RAM memory were used. Observe the cubic computing-time increase for the direct calculation case, in contrast to the logarithmic growth when the renormalization plus convolution method is used, which permits the study of electronic transport in truly macroscopic 3D lattices with multiple aperiodically located interfaces. Note also that, for short-length nanowires of 50 atoms, for example, the direct calculation represents a more efficient option than the renormalization one.

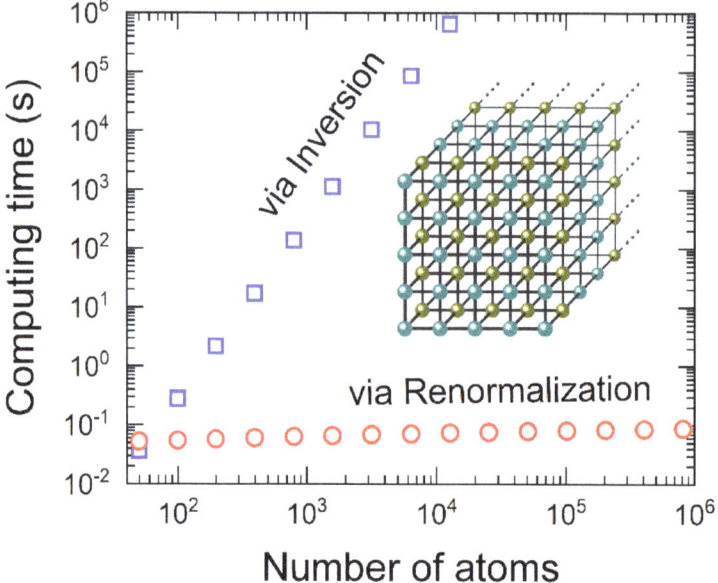

Figure 3. (color online) A log–log plot of single-energy Kubo conductivity computing time versus the total number of atoms in a quasiperiodic nanowire with a cross section of 5×5 atoms schematically illustrated in the inset, where the calculations were performed using a Fortran inversion subroutine (blue squares) and the renormalization plus convolution method of [17] (red circles).

Another widely studied 2D quasiperiodic lattice is the Penrose tiling, shown in Figure 2e, whose integrated density of states (IDOS) presents a central peak with about 10% of the total number of states separated from two symmetric bands by two finite gaps [171,172]. The presence of these gaps in macroscopic Penrose lattices has been confirmed by a real-space renormalization study [173] and analyzed by means of a square of the Hamiltonian (H^2) obtained from renormalizing one of the alternating sublattices, because the Penrose tiling is bipartite. The band center of the original Hamiltonian is mapped to the minimum eigenvalue of H^2, whose eigenfunction has antibonding symmetry and is frustrated by triangular cells in H^2 [174,175]. At the same time, the local [176] and total [177] electronic density of states in Penrose lattices were also studied by a renormalization method, neglecting the small hopping integrals corresponding to the long diagonal of kites. Similar renormalization procedures have been applied to the study of the bond percolation problem [178], phason elasticity [179], Potts spin interaction [180], critical eigenstates [181], and Hubbard model within the real-space dynamical mean-field theory [182,183].

In general, an exact RSRM for 2D Penrose lattices requires the explicit consideration of all boundary sites in each generation to calculate the next-generation Green's function, because it counts all possible paths between two arbitrary sites. This fact inhibits a suitable application of RSRM to truly macroscopic Penrose lattices, in contrast to 1D systems, where the number of boundary sites is always two. Hence, hypercubic aperiodic lattices are commonly addressed by using a renormalization plus convolution method [17].

Labyrinth lattices, shown in Figure 2g, constitute an example of non-cubically structured 2D aperiodic tiling, where a novel convolution plus renormalization method has been successfully applied [184], being the first aperiodic multidimensional lattice beyond hypercubic structures investigated by means of RSRM. This lattice was first introduced by C. Sire in 1989 obtained from a Euclidean product of two 1D aperiodic chains [185,186]. The energy spectrum of the Labyrinth tiling has been proven to be an interval if parameters λ_x and λ_y of the x and y direction chains, defined by $\lambda \equiv |t_A^2 - t_B^2|/t_A t_B$, are sufficiently close to zero, and it is a Cantor set of zero Lebesgue measure if λ_x and λ_y are large enough [187,188]. The wave packet dynamics [189] and quantum diffusion [190] in the Labyrinth tiling were also analyzed using RSRM. Labyrinth lattices based on silver-means quasiperiodic chains have been observed in a surface-wave experiment [191].

5. Vibrational Excitations

A solid of N atoms has $3N$ degrees of freedom and it can translate or rotate as a whole, hence it may have $3N - 6$ normal modes of vibration, in which all atoms move sinusoidally with the same frequency and a fixed phase relation [3,192]. The quantum of these normal vibrational modes is called phonon, who has crucial participation in the Raman scattering [193], infrared (IR) spectroscopy [194], and inelastic neutron scattering [195], as well as in thermal transport [196]. These phonons, as other elementary excitations in solids, are scattered by impurities, defects, and structural interfaces, and their transport in quasiperiodic lattices has been studied since the discovery of quasicrystals. For example, the first quasiperiodic GaAs-AlAs superlattice was built in 1985 [156] and its acoustic Raman spectrum measured from the backscattering [29] was theoretically reproduced in 1988 [28]. Using RSRM, the phonon frequencies [197,198], local DOS [199,200], transmission coefficient [201], and lattice specific heat [202] in Fibonacci chains, as well as vibrational properties in Thue–Morse [202,203], period-doubling [204], Rudin–Shapiro [204], and three-component Fibonacci [205] systems, were studied. Experimental determination of phonon behavior was carried out in 1D aperiodic lattices through the third sound on a superfluid helium film [206], while in 2D Penrose tiling using quasiperiodic arrays of Josephson junctions [207], tuning forks [208], and LC electric oscillators [209], in which anharmonic effects were also analyzed.

The lattice thermal conductance (K) given by Equation (5) of [210] is calculated using the RSRM and comparatively presented in Figure 4 for periodic (165,580,142 atoms), Fibonacci (165,580,142 atoms), Thue–Morse (134,217,729 atoms), and period doubling (134,217,729 atoms) chains with a uniform mass M and restoring force constants $\alpha_A = \frac{1}{2}(\sqrt{5} - 1)\alpha$ and $\alpha_B = \alpha$ connected to two periodic leads at their ends, where $K_0 = \pi k_B \omega_0/6$ is the quantum of thermal conductance [211], $\omega_0 = \sqrt{\alpha/M}$, and $T_0 = \hbar\omega_0/k_B$. In general, the thermal conductance of aperiodic chains diminishes with the structural disorder strength and the system length, whose temperature variation $K(T)$ is consistent with those reported in [212]. The corresponding phonon transmittance spectra are shown in Figure 4a for Fibonacci, Figure 4b for period doubling, and Figure 4c for Thue–Morse chains in comparison with that of the periodic one illustrated by the dark-yellow solid lines in each of them, while a low-temperature magnification of $K(T) - T$ is exposed in Figure 4d for the mentioned chains. Observe in Figure 4d the nearly linear behavior of $K(T)$ for the periodic case whose small deviation is caused by the finite length of system, and the presence of a crossing between $K(T)$ curves of Fibonacci and Thue–Morse chains, where the higher $K(T)$ of Thue–Morse chains at low temperature is originated from its almost one transmittance around the zero vibrational frequency, as shown in Figure 4c.

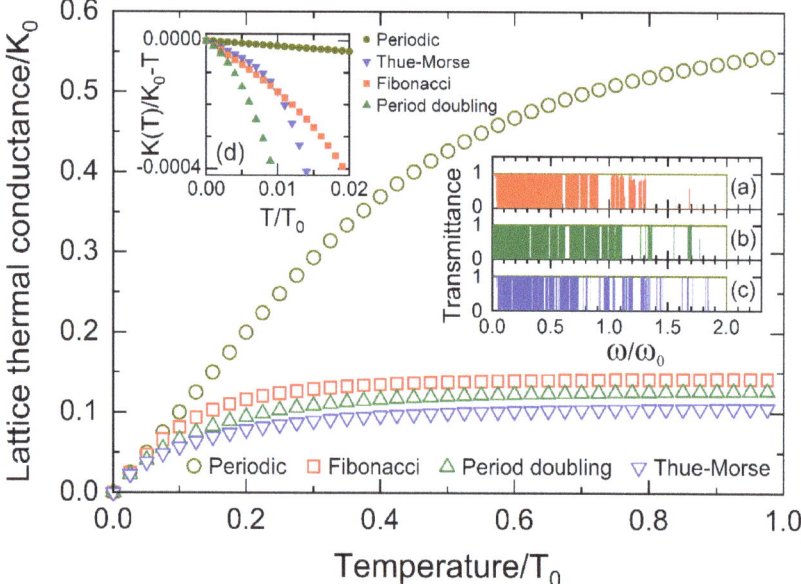

Figure 4. Lattice thermal conductance (K) as a function of temperature (T) for periodic (circles), Fibonacci (squares), period doubling (up triangles), and Thue–Morse (down triangles) chains. Insets: the corresponding phononic transmittance spectra of (**a**) Fibonacci, (**b**) period doubling, and (**c**) Thue–Morse chains, as well as (**d**) an amplification of K(T)/K$_0$ −T at the low-temperature zone.

For 3D systems, the real-space renormalization plus convolution method has been applied to the study of lattice thermal conductivity by phonons in quasiperiodic nanowires (NW), whose power–law temperature dependence as a function of the NW cross-section area has a good agreement with the experimental results [211]. The direct conversion between thermal and electrical energies can be achieved by means of thermoelectric devices, whose performance can be measured using the dimensionless figure-of-merit defined as

$$ZT = \sigma S^2 T / (\kappa_{el} + \kappa_{ph}) \tag{9}$$

where S is the Seebeck coefficient; σ is the electrical conductivity; and κ_{el} and κ_{ph} are the electronic and phononic thermal conductivities, respectively [2]. The inherent correlation between these thermoelectric quantities makes difficult to improve the value of ZT. Recently, nanowire heterostructures, such as $M_2O_3/ZnO(M = \text{In,Ga,Fe})$ with compositional segmentation, have demonstrated a significant improvement of ZT, mainly owing to the phonon scattering at composite interfaces [213]. Thermoelectricity in periodic and quasiperiodically segmented nanobelts and nanowires were comparatively studied within the Kubo–Greenwood formalism [214], and the results reveal the importance of segmentation in ZT as well as its further improvement when the quasiperiodicity is introduced, because it significantly diminishes the thermal conduction of long wavelength acoustic phonons, which are responsible for the thermal conductivity by phonons at low a temperature and do not feel local defects nor impurities [215]. Furthermore, branches of atoms attached to a nanowire may significantly modify the transport of excitations along it owing to wave interference, whose resonance produces zones of a very high value of ZT in the Hamiltonian parameter space [216]. Poly(G)-poly(C) DNA-like double chains, shown in Figure 2h, constitute another interesting example of branched low-dimensional systems, where the fishbone model and the two-site coarse grain model based on the Born potential including central and non-central interactions are used for the calculation of electrical and lattice thermal conductivities, respectively, through the Kubo–Greenwood formula [217]. The results show the appearance of gaps in phononic transmittance spectra and a remarkable enhancement of ZT when periodic interfaces between poly(G) and poly(C) segments are introduced. Such ZT can be further improved by introducing a long-range quasiperiodic order, which avoids the thermal transport of numerous low-frequency phonons responsible of the lattice thermal conduction at a low temperature. Finally, the reservoirs have an important participation on the ZT, as they constitute the boundary conditions of the quantum system and may cause resonant interferences favoring the thermoelectric transport [210].

6. Conclusions

An aperiodic solid could be thermodynamically stable by the growth of entropy, the appearance of electronic energy gap around of the Fermi energy as occurred in the Peierls instability [218], or mechanisms described by the Hume–Rothery rules [219]. Such structural asymmetry represents a singular opportunity to achieve many unique physical properties. For example, the union of positively and negatively doped semiconductors constitutes the base of current microelectronics and modern illumination. Nevertheless, the presence of these structural interfaces requires new approaches for its study and design.

From the theoretical point of view, the tight-binding or Hubbard Hamiltonian based on the Wannier functions provides an atomic scale modelling of measurable physical quantities, where the huge degrees of freedom should be efficient and accurately addressed by taking the advantage of all visible and hidden symmetries. For instance, the exciton diffusion in organic solar cells has been recently analyzed by means of an attractive Hubbard Hamiltonian and the real-space renormalization method (RSRM) [220]. For aperiodic lattices with hierarchically structured inhomogeneities, the RSRM seems to be an ideal candidate because the structural scaling rule of these aperiodic lattices can be used as the starting point of RSRM. However, this procedure is truly useful only in 1D systems because they have a constant number of boundary atoms, in contrast to multidimensional systems whose boundary-atom number grows with the system size. These boundary atoms are extremely important for the Green's function determination, that is, a precise counting of all possible paths between two arbitrary atoms. For separable Hamiltonians, such as nearest-neighbor tight-binding Hamiltonian in cubically structured aperiodic lattices, a combination of the convolution theorem and RSRM has demonstrated its effectiveness [17]. Beyond cubically structured systems, the labyrinth lattice has been the first non-hypercubic aperiodic network recently addressed by the renormalization plus convolution scheme, where a new convolution theorem for a product of Hamiltonians instead of summation in the traditional convolution theorem was developed [184]. This fact opens a new horizon for the applicability of RSRM to more complex multidimensional aperiodic structures. On the other hand, the design of electronic and optical devices based on quantum mechanical calculations has been one of the biggest dreams of physicists and engineers, and the recent advances in RSRM bring it closer because these electronic and optical devices usually contain multiple aperiodic located structural interfaces. For example, first-principle calculations have been used in the multiscale design of omnidirectional dielectric reflectors [221] and Fabry–Perot resonators [222], whose results were experimentally confirmed.

Finally, despite the proven efficiency of RSRM in the study of systems with huge degrees of freedom, there are still many challenges in the development and application of new RSRM and they might be summarized as follows: (1) extend the applicability of RSRM to multidimensional lattices with complex structural symmetry; (2) combine the RSRM with the density functional theory to address multielectron systems; and (3) apply the RSRM to strong correlated phenomena, such as the superconductivity.

Author Contributions: V.S. and C.W. both participated in the conceptualization, reference analysis, figure preparation, manuscript writing, and final revision. All authors have read and agreed to the published version of the manuscript.

Funding: This research was partially supported by the Consejo Nacional de Ciencia y Tecnología of Mexico through grant 252943 and by the National Autonomous University of Mexico (UNAM) through projects PAPIIT-IN115519 and PAPIIT-IN110020. Computations were performed at Miztli of DGTIC, UNAM.

Acknowledgments: We would like to thank Antonio Galeote for his participation at early stages of this article and Fernando Sánchez for stimulating conversations and technical assistance.

Conflicts of Interest: The authors declare no conflict of interest.

Abbreviations

The following abbreviations are used in this manuscript:

YB	Yottabytes
RSRM	Real-space renormalization method
CPA	Coherent potential approximation
CPU	Central processing unit
1D	One-dimensional
2D	Two-dimensional
3D	Three-dimensional
DC	Direct current
AC	Alternating current
DOS	Density of states
IDOS	Integrated density of states
GF	Generalized Fibonacci
TM	Thue–Morse
PD	Period doubling
NW	Nanowires
IR	Infrared
DNA	Deoxyribonucleic acid
A	Adenine
C	Cytosine
G	Guanine
T	Thymine

References

1. Chattopadhyay, D.; Queisser, H.J. Electron scattering by ionized impurities in semiconductors. *Rev. Mod. Phys.* **1981**, *53*, 745–768. [CrossRef]
2. Maciá-Barber, E. *Thermoelectric Materials: Advances and Applications*; CRC Press: Boca Raton, FL, USA, 2015; p. 77, ISBN 978-981-4463-53-9.
3. Ashcroft, N.W.; Mermin, N.D. *Solid State Physics*; Saunders College Pub.: Fort Worth, TX, USA, 1976; pp. 615–641, ISBN 0-03-083993-9.
4. Elliott, R.J.; Krumhansl, J.A.; Leath, P.L. The theory and properties of randomly disordered crystals and related physical systems. *Rev. Mod. Phys.* **1974**, *46*, 465–543. [CrossRef]
5. Shechtman, D.; Blech, I.; Gratias, D.; Cahn, J.W. Metallic phase with long-range orientational order and no translational symmetry. *Phys. Rev. Lett.* **1984**, *53*, 1951–1954. [CrossRef]
6. Janot, C. *Quasicrystals: A Primer*, 2nd ed.; Oxford University Press: Oxford, UK, 1994; pp. 223–234, ISBN 978-0-19-965740-7.
7. Kittel, C. *Introduction to Solid State Physics*, 8th ed.; John Wiley & Sons: Hoboken, NJ, USA, 2005; pp. 23–42, ISBN 0-471-41526-X.
8. Kadanoff, L.P. Scaling laws for Ising models near Tc. *Physics* **1966**, *2*, 263–272. [CrossRef]
9. Wilson, K.G. Renormalization group and critical phenomena. I. Renormalization group and the Kadanoff scaling picture. *Phys. Rev. B* **1971**, *4*, 3174–3183. [CrossRef]
10. Wilson, K.G. Renormalization group and critical phenomena. II. Phase-space cell analysis of critical behavior. *Phys. Rev. B* **1971**, *4*, 3184–3205. [CrossRef]

11. Kramer, B.; MacKinnon, A. Localization: Theory and experiment. *Rep. Prog. Phys.* **1993**, *56*, 1469–1564. [CrossRef]
12. Economou, E.N. *Green's Functions in Quantum Physics*, 3rd ed.; Springer: Berlin/Heidelberg, Germany, 2006; pp. 7–184, ISBN 978-3-540-28838-1.
13. Oviedo-Roa, R.; Pérez, L.A.; Wang, C. AC conductivity of the transparent states in Fibonacci chains. *Phys. Rev. B* **2000**, *62*, 13805–13808. [CrossRef]
14. Maciá Barber, E. *Aperiodic Structures in Condensed Matter: Fundamentals and Applications*; CRC Press: Boca Raton, FL, USA, 2009; pp. 132–153, ISBN 978-1-4200-6827-6.
15. Posamentier, A.S.; Lehmann, I. *The Fabulous Fibonacci Numbers*; Prometheus Books: New York, NY, USA, 2007; p. 26, ISBN 978-1-59102-475-0.
16. Dunlap, R.A. *The Golden Ratio and Fibonacci Numbers*; World Scientific Pub.: Singapore, 1997; pp. 35–49, ISBN 9810232640.
17. Sánchez, V.; Wang, C. Application of renormalization and convolution methods to the Kubo-Greenwood formula in multidimensional Fibonacci systems. *Phys. Rev. B* **2004**, *70*, 144207. [CrossRef]
18. Kohmoto, M.; Oono, Y. Cantor spectrum for an almost periodic Schrödinger equation and a dynamical map. *Phys. Lett. A* **1984**, *102*, 145–148. [CrossRef]
19. Kohmoto, M.; Sutherland, B.; Tang, C. Critical wave functions and a Cantor-set spectrum of a one-dimensional quasicrystal model. *Phys. Rev. B* **1987**, *35*, 1020–1033. [CrossRef] [PubMed]
20. Kohmoto, M.; Banavar, J.R. Quasiperiodic lattice: Electronic properties, phonon properties, and diffusion. *Phys. Rev. B* **1986**, *34*, 563–566. [CrossRef] [PubMed]
21. Tang, C.; Kohmoto, M. Global scaling properties of the spectrum for a quasiperiodic Schrodinger equation. *Phys. Rev. B* **1986**, *34*, 2041–2044. [CrossRef] [PubMed]
22. Kohmoto, M. Localization problem and mapping of one-dimensional wave equations in random and quasiperiodic media. *Phys. Rev. B* **1986**, *34*, 5043–5047. [CrossRef]
23. Sutherland, B.; Kohmoto, M. Resistance of a one-dimensional quasicrystal: Power-law growth. *Phys. Rev. B* **1987**, *35*, 5877–5886. [CrossRef]
24. Niu, Q.; Nori, F. Renormalization-group study of one-dimensional quasiperiodic systems. *Phys. Rev. Lett.* **1986**, *57*, 2057–2060. [CrossRef]
25. Zheng, W.M. Global scaling properties of the spectrum for the Fibonacci chains. *Phys. Rev. A* **1987**, *35*, 1467–1469. [CrossRef]
26. Roman, H.E. Hierarchical structure of a one-dimensional quasiperiodic model. *Phys. Rev. B* **1988**, *37*, 1399–1401. [CrossRef]
27. Villaseñor-González, P.; Mejía-Lira, F.; Morán-López, J.L. Renormalization group approach to the electronic spectrum of a Fibonacci chain. *Solid State Commun.* **1988**, *66*, 1127–1130. [CrossRef]
28. Wang, C.; Barrio, R.A. Theory of the Raman response in Fibonacci superlattices. *Phys. Rev. Lett.* **1988**, *61*, 191–194. [CrossRef]
29. Bajema, K.; Merlin, R. Raman scattering by acoustic phonons in Fibonacci GaAs-AlAs superlattices. *Phys. Rev. B* **1987**, *36*, 4555–4557. [CrossRef] [PubMed]
30. Chakrabarti, A.; Karmakar, S.N.; Moitra, R.K. Exact real-space renormalization-group approach for the local electronic Green's functions on an infinite Fibonacci chain. *Phys. Rev. B* **1989**, *39*, 9730–9733. [CrossRef] [PubMed]
31. Tsunetsugu, H.; Ueda, K. Ising spin system on the Fibonacci chain. *Phys. Rev. B* **1987**, *36*, 5493–5499. [CrossRef] [PubMed]
32. Ashraff, J.A.; Stinchcombe, R.B. Nonuniversal critical dynamics on the Fibonacci-chain quasicrystal. *Phys. Rev. B* **1989**, *40*, 2278–2283. [CrossRef]
33. Aldea, A.; Dulea, M. Hopping conduction on aperiodic chains. *Phys. Rev. Lett.* **1988**, *60*, 1672–1675. [CrossRef]
34. Miller, A.; Abrahams, E. Impurity conduction at low concentrations. *Phys. Rev.* **1960**, *120*, 745–755. [CrossRef]
35. López, J.C.; Naumis, G.; Aragón, J.L. Renormalization group of random Fibonacci chains. *Phys. Rev. B* **1993**, *48*, 12459–12464. [CrossRef]
36. Barrio, R.A.; Wang, C. Electron Localization in Large Fibonacci Chains. In *Quasicrystals and Incommensurate Structures in Condensed Matter*; José Yacamán, M., Romeu, D., Castaño, V., Gómez, A., Eds.; World Scientific: Singapore, 1990; pp. 448–464, ISBN 981-02-0001-3.

37. Capaz, R.B.; Koiller, B.; de Queiroz, S.L.A. Gap states and localization properties of one-dimensional Fibonacci quasicrystals. *Phys. Rev. B* **1990**, *42*, 6402–6407. [CrossRef]
38. Liu, Y.; Sritrakool, W. Branching rules of the energy spectrum of one-dimensional quasicrystals. *Phys. Rev. B* **1991**, *43*, 1110–1116. [CrossRef]
39. Chakrabarti, A.; Karmakar, S.N.; Moitra, R.K. On the nature of eigenstates of quasiperiodic lattices in one dimension. *Phys. Lett. A* **1992**, *168*, 301–304. [CrossRef]
40. Zhong, J.X.; You, J.Q.; Yan, J.R.; Yan, X.H. Local electronic properties of one-dimensional quasiperiodic systems. *Phys. Rev. B* **1991**, *43*, 13778–13781. [CrossRef] [PubMed]
41. Zhong, J.X.; Yan, J.R.; You, J.Q.; Yan, X.H. Exact renormalization-group approach for the average Green functions of aperiodic lattices. *Phys. Lett. A* **1993**, *177*, 71–75. [CrossRef]
42. Newman, M.E.J.; Stinchcombe, R.B. Hopping conductivity of the Fibonacci-chain quasicrystal. *Phys. Rev. B* **1991**, *43*, 1183–1186. [CrossRef] [PubMed]
43. Chakrabarti, A. AC-conductivity of aperiodic chains re-examined. *Z. Phys. B* **1993**, *93*, 127–131. [CrossRef]
44. Piéchon, F.; Benakli, M.; Jagannathan, A. Analytical results for scaling properties of the spectrum of the Fibonacci chain. *Phys. Rev. Lett.* **1995**, *74*, 5248–5251. [CrossRef]
45. Maciá, E.; Domínguez-Adame, F. Physical nature of critical wave functions in Fibonacci systems. *Phys. Rev. Lett.* **1996**, *76*, 2957–2960. [CrossRef]
46. Ghosh, A.; Karmakar, S.N. Electronic properties of quasiperiodic Fibonacci chain including second-neighbor hopping in the tight-binding model. *Eur. Phys. J. B* **1999**, *11*, 575–582. [CrossRef]
47. Sánchez, V.; Pérez, L.A.; Oviedo-Roa, R.; Wang, C. Renormalization approach to the Kubo formula in Fibonacci systems. *Phys. Rev. B* **2001**, *64*, 174205. [CrossRef]
48. Sánchez, V.; Wang, C. Exact results of the Kubo conductivity in macroscopic Fibonacci systems: A renormalization approach. *J. Alloys Compd.* **2002**, *342*, 410–412. [CrossRef]
49. Sánchez, V.; Wang, C. Electronic transport in quasiperiodic lattices. *J. Phys. Soc. Jpn.* **2003**, *72*, 177–178. [CrossRef]
50. Walther, D.; Baltz, R.V. Frequency dependent conductivity of Fibonacci-chains. *J. Low Temp. Phys.* **2002**, *126*, 1211–1220. [CrossRef]
51. Velhinho, M.T.; Pimentel, I.R. Lyapunov exponent for pure and random Fibonacci chains. *Phys. Rev. B* **2000**, *61*, 1043–1050. [CrossRef]
52. Wang, C.; Oviedo-Roa, R.; Pérez, L.A.; Sánchez, V. Electrical conductivity and localization in quasiperiodic lattices. *Ferroelectrics* **2001**, *250*, 305–308. [CrossRef]
53. Naumis, G.G. The stability of the renormalization group as a diagnostic tool for localization and its application to the Fibonacci case. *J. Phys. Condens. Matter* **2003**, *15*, 5969–5978. [CrossRef]
54. Vasconcelos, M.S.; Mauriz, P.W.; Albuquerque, E.L.; da Silva, E.F., Jr.; Freire, V.N. Electronic spectra of GaAs/Ga$_x$Al$_{1-x}$As superlattice with impurities arranged according to a Fibonacci sequence. *Appl. Surf. Sci.* **2004**, *234*, 33–37. [CrossRef]
55. Bakhtiari, M.R.; Vignolo, P.; Tosi, M.P. Coherent transport in linear arrays of quantum dots: The effects of period doubling and of quasi-periodicity. *Physica E* **2005**, *28*, 385–392. [CrossRef]
56. Maciá, E.; Rodríguez-Oliveros, R. Renormalization transformation of periodic and aperiodic lattices. *Phys. Rev. B* **2006**, *74*, 144202. [CrossRef]
57. Sengupta, S.; Chakrabarti, A. Wave propagation in a quasi-periodic waveguide network. *Physica E* **2005**, *28*, 28–36. [CrossRef]
58. Bakhtiari, M.R.; Vignolo, P.; Tosi, M.P. Theory of coherent transport by an ultra-cold atomic Fermi gas through linear arrays of potential wells. *Physica E* **2006**, *33*, 223–229. [CrossRef]
59. Maciá, E. Clustering resonance effects in the electronic energy spectrum of tridiagonal Fibonacci quasicrystals. *Phys. Status Solidi B* **2017**, *254*, 1700078. [CrossRef]
60. Hida, K. Quasiperiodic Hubbard Chains. *Phys. Rev. Lett.* **2001**, *86*, 1331–1334. [CrossRef] [PubMed]
61. Arredondo, Y.; Navarro, O. Electron pairing in one-dimensional quasicrystals. *Solid State Commun.* **2010**, *150*, 1313–1316. [CrossRef]
62. Hida, K. New universality class in spin-one-half Fibonacci Heisenberg chains. *Phys. Rev. Lett.* **2004**, *93*, 037205. [CrossRef] [PubMed]
63. Cassels, J.W.S. *An Introduction to Diophantine Approximation*; Cambridge University Press: Cambridge, UK, 1957; p. 133, ISBN 978-0521045872.

64. Luck, J.M.; Godreche, C.; Janner, A.; Janssen, T. The nature of the atomic surfaces of quasiperiodic self-similar structures. *J. Phys. A Math. Gen.* **1993**, *26*, 1951–1999. [CrossRef]
65. Maciá, E. Exploiting aperiodic designs in nanophotonic devices. *Rep. Prog. Phys.* **2012**, *75*, 036502. [CrossRef] [PubMed]
66. Gumbs, G.; Ali, M.K. Dynamical maps, Cantor spectra, and localization for Fibonacci and related quasiperiodic lattices. *Phys. Rev. Lett.* **1988**, *60*, 1081–1084. [CrossRef]
67. Chakrabarti, A.; Karmakar, S.N. Renormalization-group method for exact Green's functions of self-similar lattices: Application to generalized Fibonacci chains. *Phys. Rev. B* **1991**, *44*, 896–899. [CrossRef]
68. Zhong, J.X.; Yan, J.R.; You, J.Q.; Yan, X.H.; Mei, Y.P. Electronic properties of one-dimensional quasiperiodic lattices: Green's function renormalization group approach. *Z. Phys. B Condens. Matter* **1993**, *91*, 127–133. [CrossRef]
69. Zhong, J.X.; Yan, J.R.; You, J.Q. Renormalization-group approach to the local Green functions of a family of generalized Fibonacci lattices. *J. Phys. A Math. Gen.* **1991**, *24*, L949–L954. [CrossRef]
70. Yan, X.H.; Zhong, J.X.; Yan, J.R.; You, J.Q. Renormalization group of generalized Fibonacci lattices. *Phys. Rev. B* **1992**, *46*, 6071–6079. [CrossRef]
71. Yan, X.H.; You, J.Q.; Yan, J.R.; Zhong, J.X. Renormalization Group on the Aperiodic Hamiltonian. *Chin. Phys. Lett.* **1992**, *9*, 623–625.
72. Oh, G.Y.; Ryu, C.S.; Lee, M.H. Clustering properties of energy spectra for one-dimensional generalized Fibonacci lattices. *Phys. Rev. B* **1993**, *47*, 6122–6125. [CrossRef] [PubMed]
73. Zhong, J.X.; Yan, J.R.; You, J.Q. Exact Green's functions of generalized Fibonacci lattices. *J. Non Cryst. Solids* **1993**, *153*, 439–442. [CrossRef]
74. Fu, X.; Liu, Y.; Zhou, P.; Sritrakool, W. Perfect self-similarity of energy spectra and gap-labeling properties in one-dimensional Fibonacci-class quasilattices. *Phys. Rev. B* **1997**, *55*, 2882–2889. [CrossRef]
75. Walther, D.; Baltz, R.V. Path renormalization of quasiperiodic generalized Fibonacci chains. *Phys. Rev. B* **1997**, *55*, 8852–8866. [CrossRef]
76. Chakrabarti, A. The unusual electronic spectrum of an infinite quasiperiodic chain: Extended signature of all eigenstates. *J. Phys. Condens. Matter* **1994**, *6*, 2015–2024. [CrossRef]
77. Barghathi, H.; Nozadze, D.; Vojta, T. Contact process on generalized Fibonacci chains: Infinite-modulation criticality and double-log periodic oscillations. *Phys. Rev. E* **2014**, *89*, 012112. [CrossRef]
78. Wang, C.; Ramírez, C.; Sánchez, F.; Sánchez, V. Ballistic conduction in macroscopic non-periodic lattices. *Phys. Status Solidi B* **2015**, *252*, 1370–1381. [CrossRef]
79. Sánchez, F.; Sánchez, V.; Wang, C. Renormalization approach to the electronic localization and transport in macroscopic generalized Fibonacci lattices. *J. Non Cryst. Solids* **2016**, *450*, 194–208. [CrossRef]
80. Maciá, E. Spectral classification of one-dimensional binary aperiodic crystals: An algebraic approach. *Ann. Phys.* **2017**, *529*, 1700079. [CrossRef]
81. Maciá, E. The role of aperiodic order in science and technology. *Rep. Prog. Phys.* **2006**, *69*, 397–441. [CrossRef]
82. Qin, M.-G.; Ma, H.-R.; Tsai, C.-H. A renormalisation analysis of the one-dimensional Thue-Morse aperiodic chain. *J. Phys. Condens. Matter* **1990**, *2*, 1059–1072. [CrossRef]
83. Zhong, J.X.; You, J.Q.; Yan, J.R. The exact Green function of a one-dimensional Thue-Morse lattice. *J. Phys. Condens. Matter* **1992**, *4*, 5959–5965. [CrossRef]
84. Ghosh, A.; Karmakar, S.N. Trace map of a general aperiodic Thue-Morse chain: Electronic properties. *Phys. Rev. B* **1998**, *58*, 2586–2590. [CrossRef]
85. Cheng, S.-F.; Jin, G.-J. Trace map and eigenstates of a Thue-Morse chain in a general model. *Phys. Rev. B* **2002**, *65*, 134206. [CrossRef]
86. Chakrabarti, A.; Karmakar, S.N.; Moitra, R.K. Role of a new type of correlated disorder in extended electronic states in the Thue-Morse lattice. *Phys. Rev. Lett.* **1995**, *74*, 1403–1406. [CrossRef]
87. Chattopadhyay, S.; Chakrabarti, A. Role of an invariant in the existence of delocalized electronic states in generalized models of a Thue-Morse aperiodic chain. *Phys. Rev. B* **2001**, *63*, 132201. [CrossRef]
88. Maciá, E.; Domínguez-Adame, F. Exciton optical absorption in self-similar aperiodic lattices. *Phys. Rev. B* **1994**, *50*, 16856–16860. [CrossRef]
89. Hu, Y.; Tian, D.-C.; Wang, L. Renormalization group approach to the random period doubling lattice. *Phys. Lett. A* **1995**, *207*, 293–298. [CrossRef]

90. Hu, Y.; Tian, D.-C. Spectral properties of the period-doubling lattice: Exact renormalization group study. *Z. Phys. B* **1996**, *100*, 629–633. [CrossRef]
91. Lin, Z.; Kong, X.; Yang, Z.R. Critical behavior of the Gaussian model on a diamond-type hierarchical lattice with periodic and aperiodic interactions. *Phys. A* **1999**, *271*, 118–124. [CrossRef]
92. Liu, Y.; Fu, X.; Han, H.; Cheng, B.; Luan, C. Spectral structure for a class of one-dimensional three-tile quasilattices. *Phys. Rev. B* **1991**, *43*, 13240–13245. [CrossRef] [PubMed]
93. Deng, W.; Wang, S.; Liu, Y.; Zheng, D.; Zou, N. Electronic properties of a one-dimensional three-tile quasilattice. *Phys. Rev. B* **1993**, *47*, 5653–5659. [CrossRef] [PubMed]
94. Maciá, E. On the nature of electronic wave functions in one-dimensional self-similar and quasiperiodic systems. *ISRN Condens. Matter Phys.* **2014**, *2014*, 165943. [CrossRef]
95. Miroshnichenko, A.E.; Flach, S.; Kivshar, Y.S. Fano resonances in nanoscale structures. *Rev. Mod. Phys.* **2010**, *82*, 2257–2298. [CrossRef]
96. Wang, C.; González, J.E.; Sánchez, V. Enhancement of the thermoelectric figure-of-merit in nanowire superlattices. *Mater. Res. Soc. Symp. Proc.* **2015**, *1735*. [CrossRef]
97. Orellana, P.A.; Domínguez-Adame, F.; Gómez, I.; Ladrón de Guevara, M.L. Transport through a quantum wire with a side quantum-dot array. *Phys. Rev. B* **2003**, *67*, 085321. [CrossRef]
98. Miroshnichenko, A.E.; Kivshar, Y.S. Engineering Fano resonances in discrete arrays. *Phys. Rev. E* **2005**, *72*, 056611. [CrossRef]
99. Chakrabarti, A. Electronic transmission in a model quantum wire with side-coupled quasiperiodic chains: Fano resonance and related issues. *Phys. Rev. B* **2006**, *74*, 205315. [CrossRef]
100. Farchioni, R.; Grosso, G.; Parravicini, G.P. Quenching of the transmittivity of a one-dimensional binary random dimer model through side-attached atoms. *Phys. Rev. B* **2012**, *85*, 165115. [CrossRef]
101. Mardaani, M.; Rabani, H. A solvable model for electronic transport of a nanowire in the presence of effective impurities. *Superlattices Microstruct.* **2013**, *59*, 155–162. [CrossRef]
102. Pal, B. Absolutely continuous energy bands and extended electronic states in an aperiodic comb-shaped nanostructure. *Phys. Status Solidi B* **2014**, *251*, 1401–1407. [CrossRef]
103. Nandy, A.; Pal, B.; Chakrabarti, A. Tight-binding chains with off-diagonal disorder: Bands of extended electronic states induced by minimal quasi-one-dimensionality. *EPL* **2016**, *115*, 37004. [CrossRef]
104. Chakrabarti, A. Fano resonance in discrete lattice models: Controlling lineshapes with impurities. *Phys. Lett. A* **2007**, *366*, 507–512. [CrossRef]
105. Chattopadhyay, S.; Chakrabarti, A. Electronic transmission in quasiperiodic serial stub structures. *J. Phys. Condens. Matter* **2004**, *16*, 313–323. [CrossRef]
106. Nomata, A.; Horie, S. Self-similarity appearance conditions for electronic transmission probability and Landauer resistance in a Fibonacci array of T stubs. *Phys. Rev. B* **2007**, *76*, 235113. [CrossRef]
107. Ramírez, C.; Sánchez, V. Kubo conductivity of macroscopic systems with Fano defects for periodic and quasiperiodic cases by means of renormalization methods in real space. *Phys. Status Solidi A* **2013**, *210*, 2431–2438. [CrossRef]
108. Sánchez, V.; Wang, C. Resonant AC conducting spectra in quasiperiodic systems. *Int. J. Comput. Mater. Sci. Eng.* **2012**, *1*, 1250003. [CrossRef]
109. Sánchez, V.; Wang, C. Improving the ballistic AC conductivity through quantum resonance in branched nanowires. *Philos. Mag.* **2015**, *95*, 326–333. [CrossRef]
110. Lambert, C.J. Basic concepts of quantum interference and electron transport in single-molecule electronics. *Chem. Soc. Rev.* **2015**, *44*, 875–888. [CrossRef]
111. Su, T.A.; Neupane, M.; Steigerwald, M.L.; Venkataraman, L.; Nuckolls, C. Chemical principles of single-molecule electronics. *Nat. Rev. Mater.* **2016**, *1*, 16002. [CrossRef]
112. Nomata, A.; Horie, S. Fractal feature of localized electronic states in Fibonacci arrays of Aharonov-Bohm rings. *Phys. Rev. B* **2007**, *75*, 115130. [CrossRef]
113. Sil, S.; Maiti, S.K.; Chakrabarti, A. Metal-insulator transition in an aperiodic ladder network: An exact result. *Phys. Rev. Lett.* **2008**, *101*, 076803. [CrossRef] [PubMed]
114. Chakrabarti, A. Electronic transmission in bent quantum wires. *Physica E* **2010**, *42*, 1963–1967. [CrossRef]
115. Farchioni, R.; Grosso, G.; Parravicini, G.P. Electronic transmission through a ladder with a single side-attached impurity. *Eur. Phys. J. B* **2011**, *84*, 227–233. [CrossRef]

116. Pal, B.; Maiti, S.K.; Chakrabarti, A. Complete absence of localization in a family of disordered lattices. *EPL* **2013**, *102*, 17004. [CrossRef]
117. Dutta, P.; Maiti, S.K.; Karmakar, S.N. A renormalization group study of persistent current in a quasiperiodic ring. *Phys. Lett. A* **2014**, *378*, 1388–1391. [CrossRef]
118. Pal, B.; Chakrabarti, A. Engineering bands of extended electronic states in a class of topologically disordered and quasiperiodic lattices. *Phys. Lett. A* **2014**, *378*, 2782–2789. [CrossRef]
119. Pal, B.; Chakrabarti, A. Absolutely continuous energy bands in the electronic spectrum of quasiperiodic ladder networks. *Physica E* **2014**, *60*, 188–195. [CrossRef]
120. Bravi, M.; Farchioni, R.; Grosso, G.; Parravicini, G.P. Riccati equation for simulation of leads in quantum transport. *Phys. Rev. B* **2014**, *90*, 155445. [CrossRef]
121. Nandy, A.; Chakrabarti, A. Engineering flat electronic bands in quasiperiodic and fractal loop geometries. *Phys. Lett. A* **2015**, *379*, 2876–2882. [CrossRef]
122. Mukherjee, A.; Nandy, A. Spectral engineering and tunable thermoelectric behavior in a quasiperiodic ladder network. *Phys. Lett. A* **2019**, *383*, 570–577. [CrossRef]
123. Mukherjee, A.; Chakrabarti, A.; Römer, R.A. Flux-driven and geometry-controlled spin filtering for arbitrary spins in aperiodic quantum networks. *Phys. Rev. B* **2018**, *98*, 075415. [CrossRef]
124. Mukherjee, A.; Römer, R.A.; Chakrabarti, A. Spin-selective Aharonov-Casher caging in a topological quantum network. *Phys. Rev. B* **2019**, *100*, 161108. [CrossRef]
125. Chakrabarti, A. Electronic states and charge transport in a class of low dimensional structured systems. *Physica E* **2019**, *114*, 113616. [CrossRef]
126. Xu, B.; Zhang, P.; Li, X.; Tao, N. Direct conductance measurement of single DNA molecules in aqueous solution. *Nano Lett.* **2004**, *4*, 1105–1108. [CrossRef]
127. Taniguchi, M.; Kawai, T. DNA electronics. *Physica E* **2006**, *33*, 1–12. [CrossRef]
128. Sponer, J.; Leszczynski, J.; Hobza, P. Structures and energies of hydrogen-bonded DNA base pairs: A nonempirical study with inclusion of electron correlation. *J. Phys. Chem.* **1996**, *100*, 1965–1974. [CrossRef]
129. de Pablo, P.J.; Moreno-Herrero, F.; Colchero, J.; Gómez Herrero, J.; Herrero, P.; Baró, A.M.; Ordejón, P.; Soler, J.M.; Artacho, E. Absence of DC-conductivity in λ-DNA. *Phys. Rev. Lett.* **2000**, *85*, 4992–4995. [CrossRef]
130. Ladik, J.; Biczó, G.; Elek, G. Theoretical estimation of the conductivity of different periodic DNA models. *J. Chem. Phys.* **1966**, *44*, 483–485. [CrossRef]
131. Ladik, J. Energy bands in DNA. *Int. J. Quantum Chem.* **1970**, *5*, 307–317. [CrossRef]
132. Maciá, E.; Roche, S. Backbone-induced effects in the charge transport efficiency of synthetic DNA molecules. *Nanotechnology* **2006**, *17*, 3002–3007. [CrossRef]
133. Maciá, E. Electronic structure and transport properties of double-stranded Fibonacci DNA. *Phys. Rev. B* **2006**, *74*, 245105. [CrossRef]
134. Ketabi, S.A.; Khouzestani, H.F. Electronic transport through dsDNA based junction: A Fibonacci model. *Iran. J. Phys. Res.* **2014**, *14*, 67–72.
135. Joe, Y.S.; Lee, S.H.; Hedin, E.R. Electron transport through asymmetric DNA molecules. *Phys. Lett. A* **2010**, *374*, 2367–2373. [CrossRef]
136. Maciá, E. Electrical conductance in duplex DNA: Helical effects and low-frequency vibrational coupling. *Phys. Rev. B* **2007**, *76*, 245123. [CrossRef]
137. Maciá, E. π-π orbital resonance in twisting duplex DNA: Dynamical phyllotaxis and electronic structure effects. *Phys. Rev. B* **2009**, *80*, 125102. [CrossRef]
138. de Almeida, M.L.; Ourique, G.S.; Fulco, U.L.; Albuquerque, E.L.; de Moura, F.A.B.F.; Lyra, M.L. Charge transport properties of a twisted DNA molecule: A renormalization approach. *Chem. Phys.* **2016**, *478*, 48–54. [CrossRef]
139. Maciá, E. DNA-based thermoelectric devices: A theoretical prospective. *Phys. Rev. B* **2007**, *75*, 035130. [CrossRef]
140. de Moura, F.A.B.F.; Lyra, M.L.; Albuquerque, E.L. Electronic transport in poly(CG) and poly(CT) DNA segments with diluted base pairing. *J. Phys. Condens. Matter* **2008**, *20*, 075109. [CrossRef]
141. Tornow, S.; Bulla, R.; Anders, F.B.; Zwicknagl, G. Multiple-charge transfer and trapping in DNA dimers. *Phys. Rev. B* **2010**, *82*, 195106. [CrossRef]
142. Deng, C.-S.; Xu, H.; Wang, H.-Y.; Liu, X.-L. Renormalization scheme to the charge transfer efficiency of single-strand DNA with long range correlated disorder. *Mod. Phys. Lett. B* **2009**, *23*, 951–962. [CrossRef]

143. Rabani, H.; Mardaani, M. Exact analytical results on electronic transport of conjugated polymer junctions: Renormalization method. *Solid State Commun.* **2012**, *152*, 235–239. [CrossRef]
144. Liu, X.-L.; Xu, H.; Ma, S.-S.; Deng, C.-S.; Li, M.-J. Renormalization-group results of electron transport in DNA molecules with off-diagonal correlation. *Physica B* **2007**, *392*, 107–111. [CrossRef]
145. Wang, L.; Qin, Z.-J. Isolate extended state in the DNA molecular transistor with surface interaction. *Physica B* **2016**, *482*, 1–7. [CrossRef]
146. Mardaani, M.; Rabani, H. An analytical model for magnetoconductance of poly(p-phenylene)-like molecular wires in the tight-binding approach. *J. Mag. Mag. Mater.* **2013**, *331*, 28–32. [CrossRef]
147. Sarmento, R.G.; Fulco, U.L.; Albuquerque, E.L.; Caetano, E.W.S.; Freire, V.N. A renormalization approach to describe charge transport in quasiperiodic dangling backbone ladder (DBL)-DNA molecules. *Phys. Lett. A* **2011**, *375*, 3993–3996. [CrossRef]
148. Ojeda, J.H.; Pacheco, M.; Rosales, L.; Orellana, P.A. Current and Shot noise in DNA chains. *Org. Electron.* **2012**, *13*, 1420–1429. [CrossRef]
149. Pal, T.; Sadhukhan, P.; Bhattacharjee, S.M. Renormalization group limit cycle for three-stranded DNA. *Phys. Rev. Lett.* **2013**, *110*, 028105. [CrossRef]
150. Maji, J.; Bhattacharjee, S.M. Efimov effect of triple-stranded DNA: Real-space renormalization group and zeros of the partition function. *Phys. Rev. E* **2012**, *86*, 041147. [CrossRef]
151. Albuquerque, E.L.; Fulco, U.L.; Freire, V.N.; Caetano, E.W.S.; Lyra, M.L.; de Moura, F.A.B.F. DNA-based nanobiostructured devices: The role of quasiperiodicity and correlation effects. *Phys. Rep.* **2014**, *535*, 139–209. [CrossRef]
152. Lambropoulos, K.; Simserides, C. Tight-binding modeling of nucleic acid sequences: Interplay between various types of order or disorder and charge transport. *Symmetry* **2019**, *11*, 968. [CrossRef]
153. Lifshitz, R. The square Fibonacci tiling. *J. Alloys Compd.* **2002**, *342*, 186–190. [CrossRef]
154. Yang, X.-B.; Liu, Y.-Y. Electronic energy spectrum structure of the two-dimensional Fibonacci quasilattices with three kinds of atoms and one kind of bond length. *Acta Phys. Sin.* **1995**, *4*, 510–522.
155. Ma, H.-R.; Tsai, C.-H. On the energy spectra of one-dimensional quasi-periodic systems. *J. Phys. C Solid State Phys.* **1988**, *21*, 4311–4324. [CrossRef]
156. Merlin, R.; Bajema, K.; Clarke, R.; Juang, F.-Y.; Bhattacharya, P.K. Quasiperiodic GaAs-AIAs heterostructures. *Phys. Rev. Lett.* **1985**, *55*, 1768–1770. [CrossRef]
157. Fu, X.; Liu, Y.-Y.; Cheng, B.; Zheng, D. Spectral structure of two-dimensional Fibonacci quasilattices. *Phys. Rev. B* **1991**, *43*, 10808–10814. [CrossRef]
158. Yang, X.-B.; Liu, Y.-Y. Splitting rules for spectra of two-dimensional Fibonacci quasilattices. *Phys. Rev. B* **1997**, *56*, 8054–8059. [CrossRef]
159. Yang, X.-B.; Xing, D. Splitting rules for the electronic spectra of two-dimensional Fibonacci-class quasicrystals with one kind of atom and two bond lengths. *Phys. Rev. B* **2002**, *65*, 134205. [CrossRef]
160. Ashraff, J.A.; Luck, J.-M.; Stinchcombe, R.B. Dynamical properties of two-dimensional quasicrystals. *Phys. Rev. B* **1990**, *41*, 4314–4329. [CrossRef]
161. Fu, X.; Liu, Y.-Y. Renormalization-group approach for the local density of states of two-dimensional Fibonacci quasilattices. *Phys. Rev. B* **1993**, *47*, 3026–3030. [CrossRef] [PubMed]
162. Sánchez, V.; Wang, C. Kubo conductivity in two-dimensional Fibonacci lattices. *J. Non Cryst. Solids* **2003**, *329*, 151–154. [CrossRef]
163. Sánchez, V.; Wang, C. Convolution and renormalization techniques applied to the Kubo conductivity in quasiperiodic systems. *J. Non Cryst. Solids* **2004**, *345*, 518–522. [CrossRef]
164. Wang, C.; Sánchez, V.; Salazar, F. Fractal quantization of the electrical conductance in quasiperiodic systems. *Ferroelectrics* **2004**, *305*, 261–264. [CrossRef]
165. Sánchez, V.; Wang, C. Electronic transport in multidimensional Fibonacci lattices. *Philos. Mag.* **2006**, *86*, 765–771. [CrossRef]
166. Sánchez, V.; Wang, C. Renormalization-convolution approach to the electronic transport in two-dimensional aperiodic lattices. *Surf. Sci.* **2006**, *600*, 3898–3900. [CrossRef]
167. Sánchez, V.; Wang, C. A real-space renormalization approach to the Kubo–Greenwood formula in mirror Fibonacci systems. *J. Phys. A Math. Gen.* **2006**, *39*, 8173–8182. [CrossRef]
168. Sánchez, V.; Sánchez, F.; Ramírez, C.; Wang, C. Non-perturbative analysis of impurity effects on the Kubo conductivity of nano to macroscopic structures. *MRS Adv.* **2016**, *1*, 1779–1784. [CrossRef]

169. Sánchez, V. Renormalization approach to the electrical conductivity of quasiperiodic systems with defects. *Comput. Mater. Sci.* **2008**, *44*, 32–35. [CrossRef]
170. Sánchez, V.; Ramírez, C.; Sánchez, F.; Wang, C. Non-perturbative study of impurity effects on the Kubo conductivity in macroscopic periodic and quasiperiodic lattices. *Physica B* **2014**, *449*, 121–128. [CrossRef]
171. Kohmoto, M.; Sutherland, B. Electronic states on a Penrose lattice. *Phys. Rev. Lett.* **1986**, *56*, 2740–2743. [CrossRef] [PubMed]
172. Kohmoto, M.; Sutherland, B. Electronic and vibrational modes on a Penrose lattice: Localized states and band structure. *Phys. Rev. B* **1986**, *34*, 3849–3853. [CrossRef] [PubMed]
173. Wang, C.; Barrio, R.A. The electronic band structure of Penrose lattices: A renormalization approach. In *Surface Science*; Ponce, F.A., Cardona, M., Eds.; Springer Proceedings in Physics: Berlin/Heidelberg, Germany, 1991; Volume 62, pp. 67–70, ISBN 978-3-642-76378-6.
174. Barrio, R.A.; Wang, C. Some physical inferences from the quasicrystalline topology of Penrose lattices. *J. Non Cryst. Solids* **1993**, *153*, 375–379. [CrossRef]
175. Naumis, G.G.; Barrio, R.A.; Wang, C. Effects of frustration and localization of states in the Penrose lattice. *Phys. Rev. B* **1994**, *50*, 9834–9842. [CrossRef]
176. You, J.Q.; Yan, J.R.; Zhong, J.X.; Yan, X.H. Local electronic properties of two-dimensional Penrose tilings: A renormalization-group approach. *Phys. Rev. B* **1992**, *45*, 7690–7696. [CrossRef]
177. You, J.Q.; Nori, F. The real-space renormalization group and generating function for Penrose lattices. *J. Phys. Condens. Matter* **1993**, *5*, 9431–9438. [CrossRef]
178. Aoyama, H.; Odagaki, T. Bond percolation in two-dimensional quasi-lattices. *J. Phys. A Math. Gen.* **1987**, *20*, 4985–4993. [CrossRef]
179. Tang, L.-H.; Jaric, M.V. Equilibrium quasicrystal phase of a Penrose tiling model. *Phys. Rev. B* **1990**, *41*, 4524–4546. [CrossRef]
180. Xiong, G.; Zhang, Z.-H.; Tian, D.-C. Real-space renormalization group approach to the Potts model on the two-dimensional Penrose tiling. *Phys. A* **1999**, *265*, 547–556. [CrossRef]
181. Macé, N.; Jagannathan, A.; Kalugin, P.; Mosseri, R.; Piéchon, F. Critical eigenstates and their properties in one- and two-dimensional quasicrystals. *Phys. Rev. B* **2017**, *96*, 045138. [CrossRef]
182. Takemori, N.; Koga, A. Local electron correlations in a two-dimensional Hubbard model on the Penrose lattice. *J. Phys. Soc. Jpn.* **2015**, *84*, 023701. [CrossRef]
183. Takemori, N.; Koga, A. DMFT study of the local correlation effects in quasi-periodic system. *J. Phys. Conf. Ser.* **2015**, *592*, 012038. [CrossRef]
184. Sánchez, F.; Sánchez, V.; Wang, C. Ballistic transport in aperiodic Labyrinth tiling proven through a new convolution theorem. *Eur. Phys. J. B* **2018**, *91*, 132. [CrossRef]
185. Sire, C.; Mosseri, R.; Sadoc, J.-F. Geometric study of a 2D tiling related to the octagonal quasiperiodic tiling. *J. Phys. Fr.* **1989**, *50*, 3463–3476. [CrossRef]
186. Sire, C. Electronic spectrum of a 2D quasi-crystal related to the octagonal quasi-periodic tiling. *Europhys. Lett.* **1989**, *10*, 483–488. [CrossRef]
187. Takahashi, Y. Quantum and spectral properties of the Labyrinth model. *J. Math. Phys.* **2016**, *57*, 063506. [CrossRef]
188. Takahashi, Y. Products of two Cantor sets. *Nonlinearity* **2017**, *30*, 2114–2137. [CrossRef]
189. Thiem, S.; Schreiber, M. Renormalization group approach for the wave packet dynamics in golden-mean and silver-mean labyrinth tilings. *Phys. Rev. B* **2012**, *85*, 224205. [CrossRef]
190. Thiem, S.; Schreiber, M. Wavefunctions, quantum diffusion, and scaling exponents in golden-mean quasiperiodic tilings. *J. Phys. Condens. Matter* **2013**, *25*, 075503. [CrossRef]
191. Torres, M.; Adrados, J.P.; Aragón, J.L.; Cobo, P.; Tehuacanero, S. Quasiperiodic Bloch-like states in a surface-wave experiment. *Phys. Rev. Lett.* **2003**, *90*, 114501. [CrossRef]
192. Callaway, J. *Quantum Theory of Solid State*; Academic Press: New York, NY, USA, 1974; pp. 19–24, ISBN 0-12-155201-2.
193. Alfaro, P.; Cisneros, R.; Bizarro, M.; Cruz-Irisson, M.; Wang, C. Raman scattering by confined optical phonons in Si and Ge nanostructures. *Nanoscale* **2011**, *3*, 1246–1251. [CrossRef] [PubMed]
194. Atkins, P.; de Paula, J. *Physical Chemistry*, 8th ed.; W. H. Freeman and Co.: New York, NY, USA, 2006; pp. 460–468, ISBN 0-7167-8759-8.
195. Quilichini, M.; Janssen, T. Phonon excitations in quasicrystals. *Rev. Mod. Phys.* **1997**, *69*, 277–314. [CrossRef]

196. Maciá, E. Thermal conductivity and critical modes in one-dimensional Fibonacci quasicrystals. *Mater. Sci. Eng.* **2000**, *294*, 719–722. [CrossRef]
197. Chen, B.; Gong, C.-D. The properties of one-dimensional quasiperiodic lattice's phonon spectrum. *Z. Phys. B Condens. Matter* **1987**, *69*, 103–109. [CrossRef]
198. You, J.Q.; Yang, Q.B.; Yan, J.R. Phonon properties of a class of one-dimensional quasiperiodic systems. *Phys. Rev. B* **1994**, *41*, 7491–7496. [CrossRef]
199. Zhong, J.X.; Yan, J.R.; Yan, X.H.; You, J.Q. Local phonon properties of the Fibonacci-chain quasicrystal. *J. Phys. Condens. Matter* **1991**, *3*, 5685–5691. [CrossRef]
200. Yan, X.H.; Yan, J.R.; Zhong, J.X.; You, J.Q.; Mei, Y.P. An exact renormalization-group approach for local phonon properties of single-atom and double-atom generalized Fibonacci systems. *Z. Phys. B* **1993**, *91*, 467–474. [CrossRef]
201. Maciá, E. Thermal conductivity of one-dimensional Fibonacci quasicrystals. *Phys. Rev. B* **2000**, *61*, 6645–6653. [CrossRef]
202. Gumbs, G.; Dubey, G.S.; Salman, A.; Mahmoud, B.S.; Huang, D. Statistical and transport properties of quasiperiodic layered structures: Thue-Morse and Fibonacci. *Phys. Rev. B* **1995**, *52*, 210–219. [CrossRef]
203. Ghosh, A.; Karmakar, S.N. Vibrational properties of a general aperiodic Thue-Morse lattice: Role of the pseudoinvariant of the trace map. *Phys. Rev. B* **2000**, *61*, 1051–1058. [CrossRef]
204. Kroon, L.; Riklund, R. Renormalization of aperiodic model lattices: Spectral properties. *J. Phys. A Math. Gen.* **2003**, *36*, 4519–4532. [CrossRef]
205. Ghosh, A. Dynamical properties of three component Fibonacci quasicrystal. *Eur. Phys. J. B* **2001**, *21*, 45–51. [CrossRef]
206. Kono, K.; Nakada, S. Resonant transmission and velocity renormalization of third sound in one-dimensional random lattices. *Phys. Rev. Lett.* **1992**, *69*, 1185–1188. [CrossRef] [PubMed]
207. Springer, K.N.; Van Harlingen, D.J. Resistive transition and magnetic field response of a Penrose-tile array of weakly coupled superconductor islands. *Phys. Rev. B* **1987**, *36*, 7273–7276. [CrossRef]
208. He, S.; Maynard, J.D. Eigenvalue spectrum, density of states, and eigenfunctions in a two-dimensional quasicrystal. *Phys. Rev. Lett.* **1989**, *62*, 1888–1891. [CrossRef]
209. Wang, C.; Fuentes, R.; Navarro, O.; Barrio, R.A.; Barrera, R.G. Wave behavior in anharmonic Penrose lattices. *J. Non Cryst. Solids* **1993**, *153*, 586–590. [CrossRef]
210. González, J.E.; Cruz-Irisson, M.; Sánchez, V.; Wang, C. Thermoelectric transport in poly(G)-poly(C) double chains. *J. Phys. Chem. Solids* **2020**, *136*, 109136. [CrossRef]
211. Wang, C.; Salazar, F.; Sánchez, V. Renormalization plus convolution method for atomic-scale modeling of electrical and thermal transport in nanowires. *Nano Lett.* **2008**, *8*, 4205–4209. [CrossRef]
212. Zhang, Y.-M.; Xu, C.-H.; Xiong, S.-J. Phonon transmission and thermal conductance in Fibonacci wire at low temperature. *Chin. Phys. Lett.* **2007**, *24*, 1017–1020.
213. Andrews, S.C.; Fardy, M.A.; Moore, M.C.; Aloni, S.; Zhang, M.; Radmilovic, V.; Yang, P. Atomic-level control of the thermoelectric properties in polytypoid nanowires. *Chem. Sci.* **2011**, *2*, 706–714. [CrossRef]
214. González, J.E.; Sánchez, V.; Wang, C. Thermoelectricity in periodic and quasiperiodically segmented nanobelts and nanowires. *MRS Adv.* **2016**, *1*, 3953–3958. [CrossRef]
215. González, J.E.; Sánchez, V.; Wang, C. Improving thermoelectric properties of nanowires through inhomogeneity. *J. Electron. Mater.* **2017**, *46*, 2724–2736. [CrossRef]
216. Sánchez, F.; Amador-Bedolla, C.; Sánchez, V.; Wang, C. Quasiperiodic branches in the thermoelectricity of nanowires. *J. Electron. Mater.* **2019**, *48*, 5099–5110. [CrossRef]
217. González, J.E.; Sánchez, V.; Wang, C. Resonant thermoelectric transport in atomic chains with Fano defects. *MRS Commun.* **2018**, *8*, 248–256. [CrossRef]
218. Sutton, A.P. *Electronic Structure of Materials*; Oxford University Press: New York, NY, USA, 1993; pp. 107–109, ISBN 0-19-851755-6.
219. Mizutani, U. *Hume-Rothery Rules for Structurally Complex Alloy Phases*; CRC Press: Boca Raton, FL, USA, 2011; pp. 1–2, ISBN 978-1-4200-9059-8.
220. Sánchez, F.; Amador-Bedolla, C.; Sánchez, V.; Wang, C. On the role of driving force in molecular photocells. *Phys. B Phys. Condens. Matter* **2020**, *583*, 412052. [CrossRef]

221. Palavicini, A.; Wang, C. Ab-initio determination of porous silicon refractive index confirmed by infrared transmittance measurements of an omnidirectional multilayer reflector. *Appl. Phys. B* **2018**, *124*, 65. [CrossRef]
222. Palavicini, A.; Wang, C. Ab initio design and experimental confirmation of Fabry–Perot cavities based on freestanding porous silicon multilayers. *J. Mater. Sci. Mater. Electron.* **2020**, *31*, 60–64. [CrossRef]

© 2020 by the authors. Licensee MDPI, Basel, Switzerland. This article is an open access article distributed under the terms and conditions of the Creative Commons Attribution (CC BY) license (http://creativecommons.org/licenses/by/4.0/).

Review

Localization Properties of Non-Periodic Electrical Transmission Lines

Edmundo Lazo

Departamento de Física, Facultad de Ciencias, Universidad de Tarapacá, Arica, Box 6-D, Chile; edmundolazon@gmail.com or elazo@academicos.uta.cl

Received: 7 September 2019; Accepted: 1 October 2019; Published: 9 October 2019

Abstract: The properties of localization of the $I(\omega)$ electric current function in non-periodic electrical transmission lines have been intensively studied in the last decade. The electric components have been distributed in several forms: (a) aperiodic, including self-similar sequences (Fibonacci and m-tuplingtupling Thue–Morse), (b) incommensurate sequences (Aubry–André and Soukoulis–Economou), and (c) long-range correlated sequences (binary discrete and continuous). The localization properties of the transmission lines were measured using typical diagnostic tools of quantum mechanics like normalized localization length, transmission coefficient, average overlap amplitude, etc. As a result, it has been shown that the localization properties of the classic electric transmission lines are similar to the one-dimensional tight-binding quantum model, but also features some differences. Hence, it is worthwhile to continue investigating disordered transmission lines. To explore new localization behaviors, we are now studying two different problems, namely the model of interacting hanging cells (consisting of a finite number of dual or direct cells hanging in random positions in the transmission line), and the parity-time symmetry problem ($\mathcal{PT}$-symmetry), where resistances R_n are distributed according to gain-loss sequence ($R_{2n} = +R$, $R_{2n-1} = -R$). This review presents some of the most important results on the localization behavior of the $I(\omega)$ electric current function, in dual, direct, and mixed classic transmission lines, when the electrical components are distributed non-periodically.

Keywords: non-periodic systems; localization properties; electrical transmission lines

---

1. Introduction

Disordered one-dimensional quantum systems have been studied intensively since the pioneering work of Anderson [1]. It has been discovered that for one-dimensional uncorrelated disordered (random) systems, all states become localized states at the thermodynamic limit. Conversely, in periodic systems, all states are extended states, but for short-range or long-range correlated disorder, it is possible to find discrete sets or bands of extended states, respectively [2–29]. Also, these results have been verified experimentally [30–35]. In addition to correlated disordered systems, the fundamental properties of aperiodic systems have been extensively studied [36–74]. Aperiodic systems are formed by incommensurate systems and self-similar systems, and aperiodic incommensurate systems are generated by two superimposed periodic structures with incommensurate periods. The origin of incommensurability may be structural or dynamical. In the first case, there are two or more superimposed periodic structures whose periods are incommensurate, and in the second case one periodicity is related to the crystalline structure and the other to the behavior of elementary excitations that propagate through the crystal. On the other hand, self-similar systems are generated by specific substitutional rules.

After systematic studies of their properties, aperiodic systems can be classified based on two aspects: the spectral measures of their lattice Fourier transform and their Hamiltonian energy spectrum.

According to the Lebesgue's decomposition theorem, the energy spectrum of any measure in $\mathcal{R}^n$ can be uniquely decomposed into three types of spectral measures, namely (a) purely point spectra (μ_p), (b) absolutely continuous spectra (μ_{ac}) and (c) singularly continuous spectra (μ_{sc}). In addition, a combination of these measures is possible. Using spectral measures, Maciá [71,73] introduced a classification chart which includes periodic, amorphous and aperiodic systems. In this chart the abscissa is represented by the lattice Fourier transform and the ordinate is represented by the energy spectrum. In particular, we can see that the Fibonacci and the Thue–Morse systems share the same kind of singular continuous energy spectrum, known as a critical state. In this state, the wave function amplitude presents strong spatial fluctuations; however, the decaying envelope of the local maxima cannot be fitted to an exponential function, like the exponentially localized functions. On the other side, the spectral measure of the Fourier transform is different for these two self-similar systems, namely purely point spectra (μ_p) for Fibonacci systems, but singularly continuous spectra (μ_{sc}) for Thue–Morse systems. This way, the Fibonacci systems can be classified as quasi-periodic and the Thue–Morse systems are classified as aperiodic but not quasi-periodic. Despite this fundamental difference, most self-similar systems present an infinite number of gaps and consequently, the integrated density of states shows a fractal behavior. Also, the total bandwidth goes to zero in the thermodynamic limit $N \to \infty$.

This review presents recent results about the influence of the disordered distribution of electric components (capacitances and inductances) in the localization properties of dual, direct and mixed classical transmission lines (TL) [75–89]. To study the localization behavior of these non-periodic systems, the electric components have been distributed in a variety of forms: (a) aperiodic, including self-similar sequences (Fibonacci and m−tupling Thue–Morse), (b) incommensurate sequences (Aubry–André and Soukoulis–Economou), and (c) long-range correlated sequences (binary discrete and continuous). Although we are studying classical systems, the localization properties of the transmission lines are measured using the typical tools used in quantum mechanics to characterize the localization behavior of disordered systems. Specifically, we use the normalized localization length $\Lambda(\omega)$, the inverse participation ratio $IPR(\omega)$, the transmission coefficient $T(\omega)$, the global density of states $DOS(\omega)$, the average overlap amplitude C_ω, etc. Our studies indicate that the localization behavior of the classical electric transmission lines is similar to the one-dimensional tight-binding quantum model, but also displays some significant differences. Therefore it is important to keep investigating this type of classical disordered systems.

This review proceeds as follows. Section 2 describes the three ways to build classic electric transmission lines: dual, direct and mixed. Also, the allowed frequency spectrum for each kind of transmission line is calculated and, at the same time, the methods used to obtain the localization properties of these systems are described. Section 3 presents the localization behavior of transmission lines with different kinds of disorder, like aperiodic disorder and long-range correlated disorder. Section 4 shows the main results obtained so far in relation to the localization behavior of non-periodic electrical transmission lines. Also, a possible application to the study of electrical communication between neurons in included. Finally, two new lines of research to study the effect of the disorder on the localization properties of the electric current function are indicated.

2. Electric Transmission Lines

2.1. Direct and Dual Transmission Lines

We analyze ideal classical electric transmission lines considering three possible configurations, i.e., dual, direct, and mixed. We introduce the non-periodic disorder through the values of the inductances and capacitances of each cell of disordered TL [75–89].

Figure 1 shows a segment of a transmission line (dual or direct), with horizontal impedances denoted by Z_n and vertical impedances labeled γ_n. For direct TL the impedances are $Z_n = (i\omega L_n)$ and $\gamma_n = (i\omega C_n)^{-1}$, but for dual TL we have $Z_n = (i\omega C_n)^{-1}$ and $\gamma_n = (i\omega L_n)$. Here, C_n and L_n denote the capacitance and inductance values in cell n, respectively. To study the localization properties of the electric transmission lines, capacitances C_n, inductances L_n, or both are distributed using aperiodic sequences. Applying Kirchhoff's Loop Rule to three successive unit cells of the ideal TL shown in Figure 1, we obtain a linear relationship between the electric currents circulating in the $(n-1)$th, nth and $(n+1)$th cells.

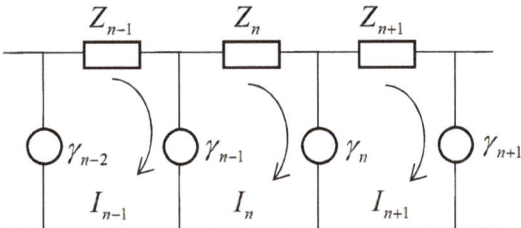

Figure 1. A partial view of an ideal transmission line. Z_n (γ_n) represent horizontal (vertical) impedances, respectively. For direct TL, Z_n is associated with inductances and γ_n with capacitances. Conversely, for dual TL, Z_n is associated with capacitances and γ_n with inductances. The arrows indicate the direction of the electric current in each cell. We arbitrarily consider the initial flow from the left, because we are using open boundary conditions

Specifically, for the direct transmission line, we find

$$\left(C_{n-1}^{-1} + C_n^{-1} - \omega^2 L_n\right) I_n - C_{n-1}^{-1} I_{n-1} - C_n^{-1} I_{n+1} = 0 \tag{1}$$

The corresponding equation for the dual TL, can be obtained using the following substitutions $\omega \to (\omega)^{-1}$, $C_n \to (L_n)^{-1}$, $L_n \to (C_n)^{-1}$, namely

$$\left(L_{n-1} + L_n - \left(\omega^2 C_n\right)^{-1}\right) I_n - L_{n-1} I_{n-1} - L_n I_{n+1} = 0 \tag{2}$$

In both cases, Equations (1) and (2) can be put in a simple generic form

$$D_n I_n - B_{n-1} I_{n-1} - B_n I_{n+1} = 0 \tag{3}$$

where

$$D_n = (B_{n-1} + B_n - A_n) \tag{4}$$

Notice that A_n always depends on frequency ω and the values of capacitances C_n or inductances L_n, while B_n only depends on C_n or L_n. To be specific, for direct TL we have $A_n = \omega^2 L_n$ and $B_n = C_n^{-1}$, but for dual TL we have $A_n = (\omega^2 C_n)^{-1}$ and $B_n = L_n$. Please note that when we introduce disorder in the off-diagonal terms, this disorder simultaneously appears in the diagonal term D_n.

2.2. Mixed Transmission Lines

The spectrum of allowed frequencies of periodic dual and direct transmission lines contains a single band. To obtain a frequency spectrum with more bands, namely a frequency selector, recently a combination of dual and direct cells, called mixed transmission line, has been studied. [87,88]. These electric systems are formed by a basic unit of $d = (p+q)$ cells consisting of a set of p successive direct cells followed by q successive dual cells. The N total number of cells in the mixed TL is given by $N = d\,N_s$, where N_s is the number of times we repeat the basic unit. Figure 2 shows a segment of a mixed TL with $p = 2$, $q = 3$ and $d = 5$. Applying Kirchhoff's Loop Rule to this system, we find a set of equations similar to Equations (1) and (2). Consequently, equations describing mixed transmission lines can also be written in the same form as the generic Equation (3), but now both coefficients D_n and B_n usually depend on the frequency ω, the parameters p, q and capacitances $C_{n,x}$, $C_{n,y}$, and inductances

$L_{n,x}$, $L_{n,y}$, corresponding to direct and dual cells, respectively. Consequently, mixed TL contain a richer parameter space to study the localization properties of the $I(\omega)$ electric current function. This allows obtaining exactly $d = (p + q)$ bands separated by gaps, containing extended and localized states and even gaps. In this way, the mixed transmission line becomes a frequency selector.

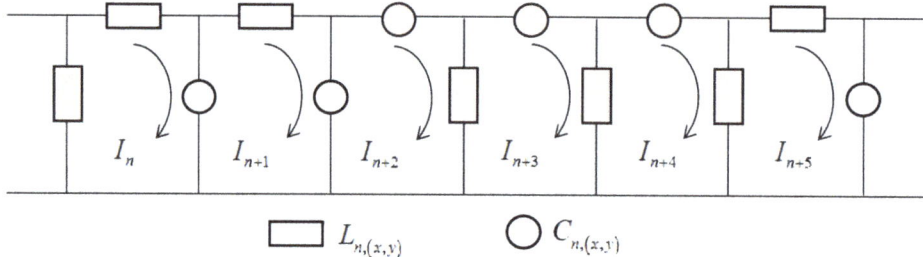

Figure 2. A segment of a mixed transmission line formed by $p = 2$ direct cells and $q = 3$ dual cells. The full system is formed by the repetition of the basic unit formed by $d = (p + q)$ cells. Inductances are represented by rectangles and capacitances by circles. In addition, dual cells are marked with orange color-filled symbols. The arrows indicate the direction of the electric current in each cell.

2.3. Relation with the Tight-Binding Model

The generic Equation (3) describing the relationship between three consecutive electric current amplitudes $I_{n-1}(\omega)$, $I_n(\omega)$ and $I_{n+1}(\omega)$ in the classical electrical transmission lines, has the same form that the equation describing the relationship between three consecutive amplitudes $\phi_{n-1}(E)$, $\phi_n(E)$ and $\phi_{n+1}(E)$ of the wave function of the one-dimensional tight-binding quantum model. This correspondence has allowed to test the effects of the disorder in one-dimensional quantum systems using classical electrical circuits with random distribution of capacitances and inductances. Consequently, Equation (3) can be mapped to the quantum one-dimensional tight-binding model

$$(E - \varepsilon_n)\phi_n - V_{n-1}\phi_{n-1} - V_n\phi_{n+1} = 0 \tag{5}$$

where ε_n is the site energy, V_n the hopping between neighboring sites, E the eigenenergy, and ϕ_n is the eigenfunction. In this quantum model, it is always possible to study separately the diagonal disordered case and the off-diagonal disordered case. However, in classical electrical TL, indicated by relations (1) and (2), the introduction of disorder in the off-diagonal terms necessarily implies that the disorder appears in both the diagonal and off-diagonal terms. Nonetheless, a correspondence between the tight-binding Equation (5) and Equations (1) and (2) exists. Applying the following transformation in the tight-binding Equation (5) we obtain the direct TL (1),

$$E = -\omega^2, \quad \varepsilon_n = -L_n^{-1}\left(C_{n-1}^{-1} + C_n^{-1}\right), \quad \phi_n = I_n (L_n)^{-\frac{1}{2}} \tag{6}$$

$$V_{n-1} = C_{n-1}^{-1}(L_{n-1}L_n)^{-\frac{1}{2}}, \quad V_n = C_n^{-1}(L_nL_{n+1})^{-\frac{1}{2}} \tag{7}$$

To obtain the dual TL (2) from the tight-binding equation, it suffices to do the following changes in transformations (6) and (7), that is $\omega \to (\omega)^{-1}$, $C_n \to (L_n)^{-1}$, $L_n \to (C_n)^{-1}$. This correspondence between the tight-binding quantum model and the classical transmission lines allows checking the localization behavior of quantum disordered one-dimensional systems using disordered TL.

2.4. Spectrum of Allowed Frequencies

To obtain the spectrum of allowed frequencies of dual and direct TL, the generic Equation (3), $D_n I_n - B_{n-1} I_{n-1} - B_n I_{n+1} = 0$, can be written in the following form:

$$I_{n+1} = \frac{D_n}{B_n} I_n - \frac{B_{n-1}}{B_n} I_{n-1} \tag{8}$$

Considering the trivial relation $I_n = I_n$, we obtain a map in the plane (I_n, I_{n+1})

$$\begin{pmatrix} I_{n+1} \\ I_n \end{pmatrix} = M_n \begin{pmatrix} I_n \\ I_{n-1} \end{pmatrix} \tag{9}$$

where matrix M_n is given by

$$M_n = \begin{pmatrix} \frac{D_n}{B_n} & -\frac{B_{n-1}}{B_n} \\ 1 & 0 \end{pmatrix} \tag{10}$$

The trajectories of the map (9) can be used to determine the extended or localized character of the electric current function $I_n(\omega)$, i.e., for extended states the trajectories are bounded, but unbounded for localized states. Also, the λ_n eigenvalues of the M_n matrix can be written in the following complex form:

$$\lambda_n = \frac{D_n}{2B_n} \pm i\sqrt{\frac{B_{n-1}}{B_n} - \left(\frac{D_n}{2B_n}\right)^2} \tag{11}$$

For a given frequency ω, the map (9) is stable if the eigenvalues λ_n of the matrix M_n are complex numbers, which also implies that the trajectories of the map are bounded, which in turn means that the electric current $I_n(\omega)$ is an extended function. λ_n is a complex number if condition $\left(\frac{B_{n-1}}{B_n}\right) > \left(\frac{D_n}{2B_n}\right)^2$ is met. For $\left(\frac{B_{n-1}}{B_n}\right) = \left(\frac{D_n}{2B_n}\right)^2$ we find the separatrix between localized states $\left(\left(\frac{B_{n-1}}{B_n}\right) < \left(\frac{D_n}{2B_n}\right)^2\right)$ and extended states. Consequently, the spectrum of allowed frequencies for direct and dual TL is given by general condition

$$\left(\sqrt{B_n} - \sqrt{B_{n-1}}\right) < \sqrt{A_n} < \left(\sqrt{B_n} + \sqrt{B_{n-1}}\right) \tag{12}$$

The coefficients A_n and B_n depend on the type of transmission line considered, direct or dual. For direct TL we have $B_n = C_n^{-1}$ and $A_n = \omega^2 L_n$ and the allowed frequencies are given by

$$\left((L_n C_n)^{-\frac{1}{2}} - (L_n C_{n-1})^{-\frac{1}{2}}\right) < \omega < \left((L_n C_n)^{-\frac{1}{2}} + (L_n C_{n-1})^{-\frac{1}{2}}\right) \tag{13}$$

For pure direct TL with $L_n = L_0$, $C_n = C_0$, $\forall n$, we find the typical band of frequencies, i.e., $0 < \omega < \frac{2}{\sqrt{L_0 C_0}}$. Conversely, for dual TL we have $B_n = L_n$ and $A_n = (\omega^2 C_n)^{-1}$ and the allowed frequencies are

$$\left(\sqrt{C_n L_n} + \sqrt{C_n L_{n-1}}\right)^{-1} < \omega < \left(\sqrt{C_n L_n} - \sqrt{C_n L_{n-1}}\right)^{-1} \tag{14}$$

For pure dual TL with $L_n = L_0$, $C_n = C_0$, $\forall n$, we also find the typical band of frequencies, i.e., $\frac{1}{2\sqrt{L_0 C_0}} < \omega < \infty$.

Next, we determined the frequency spectrum allowed for mixed transmission lines with p direct cells and q dual cells in the basic unit. The frequency spectrum shows a set of $d = (p+q)$ allowed bands separated by gaps in a bounded region of frequencies. The size and position of these d bands depends on the set of parameters that define the mixed TL, namely p, q, and the values of capacitances $C_{n,x}$, $C_{n,y}$, and inductances $L_{n,x}$, $L_{n,y}$ in direct or dual cell, respectively. Using the generic Equation (3) we write the relationship between three consecutive cells of the mixed TL. Starting from the first site with index n belonging to the direct type cell, we write p equations. After that we write q equations corresponding to dual cells. We repeat this process until we generate the complete mixed TL. Using a matrix decimation process [90,91] we can eliminate equations with sites between $(n+1)$

and $(n+d-1)$, and between $(n+d+1)$ and $(n+2d-1)$ and so on. This process allows us to write a new generic equation with renormalized coefficients D^R_{n+jd} and B^R_{n+jd}, which connects sites that are separated by a distance d, namely

$$D^R_{n+jd} I_{n+jd} - B^R_{n+(j-1)d} I_{n+(j-1)d} - B^R_{n+jd} I_{n+(j+1)d} = 0 \qquad (15)$$

where $j = 0, 1, 2, 3, \ldots$ Studying bounded trajectories of the new renormalized map, we obtain the spectrum of allowed frequencies for mixed TL, i.e.,

$$\left| D^R_{n+jd} \right| < 2 \sqrt{B^R_{n+(j-1)d} B^R_{n+jd}} \qquad (16)$$

From this relationship, two algebraic inequations of degree d in the variable ω^2 are found. Solving both inequations, exactly $d = (p+q)$ bands are obtained, within which we can observe extended and localized states, and even gaps. The size and position of these $d = (p+q)$ bands depends on the number of direct cells p and the number of dual cells q that form the mixed TL, as well as on the values of capacitances $C_{n,x}$, $C_{n,y}$ and inductances $L_{n,x}$, $L_{n,y}$ of direct and dual cells, respectively.

2.5. Methods to Obtain the $I_n(\omega)$ Electric Current Function

The $I_n(\omega)$ amplitudes of the electric current function, are obtained solving the generic Equation (3). In this paper we only consider two methods to solve this equation: (a) the recurrence method and (b) The Hamiltonian map method.

2.5.1. Recurrence Method

The $I_n(\omega)$ electric current amplitude in each cell, can be calculated using the following method. First, the generic Equation (3) is divided by I_n and then γ_n is defined as follows

$$\gamma_n = \left(B_n \frac{I_{n+1}}{I_n} \right) \qquad (17)$$

Then, Equation (3) is transformed into a recurrence equation for γ_n,

$$\gamma_n = D_n - \frac{1}{\gamma_{n-1}} (B_{n-1})^2 \qquad (18)$$

where $2 \leq n \leq N$. Iterating this equation, and starting with $\gamma_1 = D_1 = (B_1 - A_1)$, the full set of γ_n values, with $n = 1, 2, 3, \ldots, N$ is obtained. With these γ_n values, and using an arbitrary initial amplitude value $I_1 = 1$, the full set of amplitudes $\{I_n\}$ of the electric current function, can be calculated, i.e.,

$$I_{n+1} = \left(\frac{\gamma_n}{B_n} \right) I_n \qquad (19)$$

with $1 \leq n \leq (N-1)$. After that, the electric current function is normalized, i.e., $\sum_{n=1}^{N} |I_n|^2 = 1$.

Exactly the same procedure can be applied to calculate $\{I_n\}$ for mixed transmission lines, because the coefficients γ_n are always defined from the generic Equation (3), so it is not necessary to use the renormalized generic Equation (15). Also, the same is valid for the Hamiltonian map method.

2.5.2. Hamiltonian Map Method

Starting from the generic Equation (3), $D_n I_n - B_{n-1} I_{n-1} - B_n I_{n+1} = 0$, we are building a two-dimensional map (the Hamiltonian map) [28,81,83,87,88]. From the study of this map we will obtain (a) the full set of electric current amplitudes $\{I_n\}$ (from which we will obtain the localization properties) and (b) the $T(\omega)$ transmission coefficient of the disordered TL (a crucial localization tool).

Using electric current amplitudes I_n, we define the coordinate x_n at cell n, and the momentum p_n (Hamiltonian description) in the following form:

$$x_n = I_n \tag{20}$$
$$p_n = B_{n-1}(I_n - I_{n-1})$$

The generic Equation (3) can also be written using the coordinate x_n,

$$D_n x_n - B_{n-1} x_{n-1} - B_n x_{n+1} = 0 \tag{21}$$

with $D_n = (B_{n-1} + B_n - A_n)$. After some algebra, we obtain the Hamiltonian map as a function of the coefficient A_n, α_n and β_n

$$x_{n+1} = \alpha_n x_n + \beta_n p_n \tag{22}$$
$$p_{n+1} = -A_n x_n + p_n$$

where for simplicity we have defined $\beta_n = \left(\frac{1}{B_n}\right)$ and $\alpha_n = (1 - A_n \beta_n)$. This map can be written as $Z_{n+1} = M_n Z_n$, where $Z_{n+1} = \begin{pmatrix} x_{n+1} & p_{n+1} \end{pmatrix}^\tau$ and M_n is given by

$$M_n = \begin{pmatrix} \alpha_n & \beta_n \\ -A_n & 1 \end{pmatrix} \tag{23}$$

The trajectories of this map in the phase space (x, p) determine the localization properties of the $I_n(\omega)$ electric current amplitudes, namely for bounded trajectories $I_n(\omega)$ is an extended function, but for unbounded trajectories $I_n(\omega)$ is a localized function. Importantly, the study of the map's evolution (22) at "time" n, is similar to the transfer matrix method [5,28] used to study the localization behavior of disordered systems. On the other hand, starting from this map, the spectrum of allowed frequencies we can be calculated studying the complex eigenvalues of the Hamiltonian matrix M_n (see Section 2.4).

Next, variables (x, p) of the map (22) are changed by the canonical variables (r, θ) in the following way

$$x = r \sin \theta \tag{24}$$
$$p = r \cos \theta \tag{25}$$

Using (24) and (25) the map (22) becomes

$$r_{n+1} \sin \theta_{n+1} = \alpha_n r_n \sin \theta_n + \beta_n r_n \cos \theta_n \tag{26}$$
$$r_{n+1} \cos \theta_{n+1} = -A_n r_n \sin \theta_n + r_n \cos \theta_n \tag{27}$$

Dividing Equation (26) by Equation (27) we obtain a recurrence equation from which we can calculate θ_{n+1} as a function of θ_n, namely

$$\tan \theta_{n+1} = \frac{\beta_n + \alpha_n \tan \theta_n}{1 - A_n \tan \theta_n} \tag{28}$$

Now, squaring Equations (26) and (27) and adding them together, we have

$$\left(\frac{r_{n+1}}{r_n}\right)^2 = \left(\alpha_n^2 + A_n^2\right) \sin^2 \theta_n + \left(\beta_n^2 + 1\right) \cos^2 \theta_n + (\alpha_n \beta_n - A_n) \sin 2\theta_n \tag{29}$$

Defining Γ_n as
$$\Gamma_n = \frac{r_{n+1}}{r_n} \tag{30}$$
the recurrence equation to calculate r_{n+1} as a function of r_n, is as follows
$$r_{n+1} = r_n \Gamma_n \tag{31}$$

In this way, for a fixed frequency ω, and starting with an initial condition (r_0, θ_0), the full set of values of θ_n and r_n. can be obtained. Then, using $I_n = x_n$ and $x_n = r_n \sin \theta_n$, we can calculate the following relationship
$$\frac{I_{n+1}}{I_n} = \frac{x_{n+1}}{x_n} = \frac{r_{n+1} \sin \theta_{n+1}}{r_n \sin \theta_n} = \Gamma_n \left(\frac{\sin \theta_{n+1}}{\sin \theta_n}\right) \tag{32}$$
from which we obtain the recurrence relation to calculate all electric current amplitudes I_n, as a function of Γ_n and θ_n, namely
$$I_{n+1} = \Gamma_n \left(\frac{\sin \theta_{n+1}}{\sin \theta_n}\right) I_n \tag{33}$$
where $1 \leq n \leq (N-1)$.

2.6. Diagnostic Tools

Diagnostic tools have been introduced in the literature to study disordered quantum systems, because the Bloch theorem cannot be applied in the non-periodic case. To accurately estimate the degree of localization of the quantum wave function, it is generally necessary to simultaneously apply two or more different diagnostic tools. It is important to note that these diagnostic tools also allow us to determine the localization properties of classical systems, such as harmonic chains and electric transmission lines.

2.6.1. Usual Diagnostic Tools

To study the localization behavior of the disordered electric TL as a function of the frequency ω and as a function of the kind and degree of disorder, we deploy tools used in the study of the localization behavior of quantum systems: the Lyapunov exponent $\lambda(\omega)$, the normalized localization length $\Lambda(\omega)$, the participation number $D(\omega)$, the inverse participation ratio $IPR(\omega)$, the global density of states $DOS(\omega)$ and the transmission coefficient $T(\omega)$. In addition, to characterize the localization behavior of disordered TL, we study the $R_q(\omega)$ Rényi entropies [92] and the moments $\mu_q(\omega)$. All localization tools are defined as a function of the normalized electric current amplitude $I_n(\omega)$, namely $\sum_{n=1}^{N} |I_n(\omega)|^2 = 1$. In the quantum case, the localization properties are measured using the $\phi_n(E)$ amplitude of the normalized quantum wave function.

The $\lambda(\omega)$ Lyapunov exponent is defined as
$$\lambda(\omega) = \lim_{N \to \infty} \frac{1}{N} \sum_{n=1}^{N} \ln \left|\frac{I_{n+1}}{I_n}\right| \tag{34}$$

For extended states the following condition is met: $\lambda(\omega) \leq \frac{1}{N}$. From (34) we define the $Loc(\omega)$ localization length as $Loc(\omega) = \lambda^{-1}(\omega)$. Then, the $\Lambda(\omega)$ normalized localization length is defined as
$$\Lambda(\omega) = \frac{Loc(\omega)}{N} = (N\lambda(\omega))^{-1} \tag{35}$$

For extended states we have $\Lambda(\omega) \geq 1$ and for localized states we obtain $\Lambda(\omega) < 1$.

Next, we considered the $\mu_q(\omega)$ moments of the $I_n(\omega)$ electric current function. Given that we are working with normalized electric current $\sum_{n=1}^{N} |I_n|^2 = 1$, we can define the moments $\mu_q(\omega)$ in the following form

$$\mu_q(\omega) = \sum_{n=1}^{N} |I_n|^{2q} \tag{36}$$

For homogeneous distribution of $I_n(\omega)$, i.e., for $I_n(\omega) = \frac{1}{\sqrt{N}}$, $\forall n$, we find $\mu_q(\omega) = N^{-(q-1)}$. This case corresponds to the most extended case. Conversely, for fully localized states in which $I_n(\omega) = \delta_{n,m}$, we obtain $\mu_q(\omega) = 1$. The participation number $D(\omega)$ can be defined using $\mu_2(\omega)$ namely

$$D(\omega) = \frac{1}{\mu_2} = \left(\sum_{n=1}^{N} |I_n|^4 \right)^{-1} \tag{37}$$

with $1 \leq D(\omega) \leq N$. For extended states, $D(\omega)$ scales proportional to the N system size, which implies that $\ln(D(\omega))$ versus $\ln(N)$ is a straight line with slope m approximate to $m = 1$. Also, we can define the $\xi(\omega)$ normalized participation number as $\xi(\omega) = \frac{D(\omega)}{N}$, with $\frac{1}{N} \leq \xi(\omega) \leq 1$. Consequently, for extended states, $\xi(\omega)$ tends to a constant value as a function of N. In particular, for periodic systems $\xi(\omega) = \left(\frac{2}{3}\right)$. Conversely, for localized states $\xi(\omega) \to 0$. In addition, the moment μ_2 is known as the inverse participation ratio $IPR(\omega) = \mu_2 = D^{-1}(\omega)$. Consequently, $\frac{1}{N} < IPR(\omega) \leq 1$. Sometimes it is useful to calculate $(N \times IPR)$. In this case, for localized states $(N \times IPR) \to N$ and for extended states $(N \times IPR) \approx 1$. In particular, for the most extended case $(N \times IPR) = \left(\frac{3}{2}\right)$. Notice that $\xi(\omega) = (N \times IPR)^{-1}$.

Also, some of these magnitudes can be obtained as a special case of the $R_q(\omega)$ Rényi entropies [92] defined as

$$R_q(\omega) = \frac{1}{1-q} \ln \sum_{n=1}^{N} |I_n|^{2q}, \quad q \neq 1 \tag{38}$$

In the limit $q \to 1$ we obtain the Shannon entropy ($S(\omega) = \lim_{q \to 1} R_q(\omega) = R_1$)

$$S(\omega) = -\sum_{n=1}^{N} |I_n|^2 \ln |I_n|^2 \tag{39}$$

For $q = 2$ we find $R_2(\omega) = \ln D = -\ln IPR$. Moreover, the Rényi entropies $R_q(\omega)$ can be defined using the $\mu_q(\omega)$ moments in the following form

$$R_q(\omega) = \frac{1}{1-q} \ln \mu_q(\omega), \quad q \neq 1 \tag{40}$$

2.6.2. The Average Overlap Amplitude C_ω

Another tool recently introduced in the literature is the C_{ij}^ω overlap amplitude [84,86–88]. For fixed frequency ω, this quantity measures the overlap between electric current amplitudes $I_i(\omega)$ and $I_j(\omega)$, corresponding to two cells i and j of the TL and is defined as $C_{ij}^\omega = 2|I_i(\omega) I_j(\omega)|$. For homogeneous distribution of $I_j(\omega)$, i.e., for $I_j(\omega) = \frac{1}{\sqrt{N}}$, $\forall j$, we find $C_{ij}^\omega = \frac{2}{N}$. This case corresponds to the most extended case. On the contrary, for fully localized states such as $I_j(\omega) = \delta_{i,j}$, we obtain $C_{ij}^\omega = 0$. Given that C_{ij}^ω depends on each pair of sites i and j, we define the average overlap amplitude $C_\omega = \langle C_{ij}^\omega \rangle$ considering all cells of the TL, namely

$$C_\omega = \langle C_{ij}^\omega \rangle = \frac{1}{d} \sum_{i<j} C_{ij}^\omega \tag{41}$$

where $d = \frac{1}{2}N(N-1)$. The overlap amplitude is based on the definition of quantum entanglements between a pair of qubits, i and j, called pairwise entanglement (pairwise concurrence) [93,94].

Next, we considered the power of $2q$ of the overlap amplitude, C_{ij}^{ω}, i.e.,

$$\left(C_{ij}^{\omega}\right)^{2q} = 2^{2q}\left|I_i(\omega)I_j(\omega)\right|^{2q} \tag{42}$$

The average of this quantity over all cells of the TL, is given by

$$\left\langle\left(C_{ij}^{\omega}\right)^{2q}\right\rangle = \frac{2^{2q}}{d}\sum_{i<j}|I_i(\omega)|^{2q}|I_j(\omega)|^{2q} \tag{43}$$

After some algebra [84], this expression can be written as a function of the μ_q moments (36) of the electric current function, namely

$$\left\langle\left(C_{ij}^{\omega}\right)^{2q}\right\rangle = \frac{2^{2q-1}}{d}\left((\mu_q)^2 - \mu_{2q}\right) \tag{44}$$

This relationship indicates that $\left\langle\left(C_{ij}^{\omega}\right)^{2q}\right\rangle$ can determine the localization degree for any disordered system. For the case $q = \frac{1}{2}$, we obtain a simple expression to calculate the average overlap amplitude C_{ω}, i.e.,

$$C_{\omega} = \left\langle C_{ij}^{\omega}\right\rangle = \frac{1}{d}\left\{\left(\sum_{n=1}^{N}|I_n|\right)^2 - 1\right\} \tag{45}$$

Also, for $q = 1$, $\left\langle\left(C_{ij}^{\omega}\right)^{2}\right\rangle$ can be calculated as a function of the $\xi(\omega)$ normalized participation number [84,95],

$$\left\langle\left(C_{ij}^{\omega}\right)^{2}\right\rangle = \frac{2}{d}\left\{1 - \frac{1}{N\xi(\omega)}\right\} \tag{46}$$

For localized states $\xi(\omega) \to \frac{1}{N}$, which implies that $\left\langle\left(C_{ij}^{\omega}\right)^{2}\right\rangle \to 0$. Conversely, for extended states $\xi(\omega) \to 1$, then $\left\langle\left(C_{ij}^{\omega}\right)^{2}\right\rangle = \left(\frac{2}{N}\right)^2$.

In general, for any value of q, for homogeneous extended states, the following scaling relationship is obtained

$$\left\langle\left(C_{ij}^{\omega}\right)^{2q}\right\rangle = \left(\frac{2}{N}\right)^{2q} \tag{47}$$

but for fully localized states, we find $\left\langle\left(C_{ij}^{\omega}\right)^{2q}\right\rangle = 0$. In particular, for $q = \frac{1}{2}$, we find that $\left(N\left\langle C_{ij}^{\omega}\right\rangle\right)$ scales like $(NC_{\omega}) \to 2$, which means that for extended states, (NC_{ω}) is independent of system size N.

In this way, the average overlap amplitude and its powers $\left\langle\left(C_{ij}^{\omega}\right)^{2q}\right\rangle$ can adequately determine the degree of localization of the disordered TL. Finally, $\left\langle\left(C_{ij}^{\omega}\right)^{2q}\right\rangle$ can also be expressed as a function of the Rényi entropies [84,95]

$$\left\langle\left(C_{ij}^{\omega}\right)^{2q}\right\rangle = \frac{2^{2q-1}}{d}\left\{\left(e^{R_q}\right)^{2(1-q)} - \left(e^{R_{2q}}\right)^{(1-2q)}\right\} \tag{48}$$

The results shown in this subsubsection are valid even for the quantum case, considering that C_{ij}^{ω} represents the quantum entanglements between a pair of qubits C_{ij}^{E}, i and j, called pairwise entanglement (pairwise concurrence) [93,94], i.e., $C_{ij}^E = 2 |\phi_i(E) \phi_j(E)|$, where $\phi_j(E)$ represents the amplitude of the quantum wave function for the eigenstate with eigenenergy E.

2.6.3. The Transmission Coefficient T_ω

To study the transmission properties of disordered TL using the Hamiltonian map (22), the disordered segment must be embedded in two semi-infinite ordered TL, in a similar way to the transfer matrix method [5,28]. From the Hamiltonian map formalism discussed above, the transmission coefficient $T(\omega)$ can be calculated from the expression [5,28,81,83]

$$T(\omega) = \frac{2}{1 + Z_N} \quad (49)$$

For bounded trajectories of the Hamiltonian map (22), we have $Z_N \to 1$ and $T(\omega) \to 1$. This behavior indicates that the electric current function $I(\omega)$ is an extended function. On the contrary, for unbounded trajectories we have $Z_N \to \infty$ and $T(\omega) \to 0$. In this case, $I(\omega)$ is a localized function. Z_N is defined as

$$Z_N = \frac{1}{2}\left(\left(r_N^{(1)}\right)^2 + \left(r_N^{(2)}\right)^2\right) \quad (50)$$

where $r_N^{(1,2)}$ are the radii of two trajectories at "time" $n = N$ that start from two perpendicular initial points, i.e.,

$$\begin{aligned}\left(r_0^{(1)}, \theta_0^{(1)}\right) &= (1, 0) \\ \left(r_0^{(2)}, \theta_0^{(2)}\right) &= \left(1, \frac{\pi}{2}\right)\end{aligned} \quad (51)$$

The radii $r_N^{(1,2)}$ can be calculated using the relationship (31), i.e., $r_{n+1} = r_n \Gamma_n$, then $\left(r_N^{(j)}\right)^2$ is given by

$$\left(r_N^{(j)}\right)^2 = \prod_{n=1}^{N} \left(\Gamma_n^{(j)}\right)^2, \quad j = 1, 2 \quad (52)$$

Or in another form

$$\left(r_N^{(j)}\right)^2 = \exp\left(2 \sum_{n=1}^{N} \ln\left(\Gamma_n^{(j)}\right)\right), \quad j = 1, 2 \quad (53)$$

Therefore, Z_N is given by

$$Z_N = \frac{1}{2}\left[\exp\left(2 \sum_{n=1}^{N} \ln\left(\Gamma_n^{(1)}\right)\right) + \exp\left(2 \sum_{n=1}^{N} \ln\left(\Gamma_n^{(2)}\right)\right)\right] \quad (54)$$

In this way we have a procedure to calculate the transmission coefficient $T(\omega)$.

3. Disordered Transmission Lines

In this section, we study the localization behavior of the $I(\omega)$ electric current function when we distribute capacitances and inductances in a non-periodic way in dual, direct and mixed transmission lines. Here we will consider (a) aperiodic systems formed by self-similar sequences and incommensurate sequences, and (b) long-range correlated sequences. The general results indicate that the band structure of non-periodic systems is determined by the type of transmission line (dual, direct or mixed) in which the disorder is introduced, and that the existence of discrete sets or extended state bands in the thermodynamic limit, it depends on the type of aperiodic disorder used to distribute the electrical components.

3.1. Aperiodic Transmission Lines

3.1.1. Generalized Fibonacci Sequence

The generalized Fibonacci quasi-periodic sequence is given by substitution rule $A \to A^m B^n$, $B \to A$. The corresponding substitution matrix M is given by

$$M = \begin{bmatrix} m & n \\ 1 & 0 \end{bmatrix} \tag{55}$$

where the elements of the substitution matrix indicate the number of times a given letter, A or B, appears in the substitution rule, without considering the order in which these letters occur. The number of letters that appear after applying the substitution rule j times, is given by the generalized Fibonacci numbers F_j, namely $F_j = m F_{j-1} + n F_{j-2}$, $j \geq 2$, with $F_0 = F_1 = 1$. When the number of iterations j goes to infinity, the ratio between two consecutive Fibonacci numbers F_j and F_{j-1} tends to a constant number $\sigma_{m,n}$ called the mean of incommensurability, i.e.,

$$\sigma_{m,n} = \lim_{j \to \infty} \left(\frac{F_j}{F_{j-1}} \right) = \frac{1}{2} \left(m + \sqrt{m^2 + 4n} \right) \tag{56}$$

In addition, the relative frequency of both types of letters $n_A = \frac{N_A}{N}$ and $n_B = \frac{N_B}{N}$ in the limit $j \to \infty$ is given by $n_A = \frac{\sigma_{m,n}}{\sigma_{m,n}+n}$ and $n_B = \frac{n}{\sigma_{m,n}+n}$. Please note that the mean of incommensurability $\sigma_{m,n}$ can also be obtained as the maximal eigenvalue $\lambda = \sigma_{m,n}$ of the substitution matrix M (55). For the case $m=1$ and $n=1$, we obtain the golden mean $\sigma_{1,1} = \frac{1}{2} \left(1 + \sqrt{5} \right)$, and the corresponding Fibonacci sequence is the following:

$$A \to AB \to ABA \to ABAAB \to ABAABABA \to \cdots$$

Some of the other Fibonacci means [71,73,74] which have been studied are: the Silver mean $\sigma_{2,1} = \left(1 + \sqrt{2} \right)$, the Copper mean $\sigma_{1,2} = 2$, the Bronze mean $\sigma_{3,1} = \frac{1}{2} \left(3 + \sqrt{13} \right)$, the Nickel mean $\sigma_{1,3} = \frac{1}{2} \left(1 + \sqrt{13} \right)$, etc. See Maciá [73] for a spectral classification of one-dimensional binary aperiodic crystals, studying the eigenvalues $\lambda_{\pm}$ and the determinant $|M|$ of the substitution matrix M.

The Fibonacci tight-binding quantum disordered systems have been studied exhaustively by [37,38,40–46,59,60,63,71]. For the diagonal disordered case, the global number of sub-bands is exactly four. However, in the off-diagonal disordered case, the global number of sub-bands is exactly three. In both cases, each sub-band is divided into three sub-bands until it is resolvable. This self-replication behavior is characteristic of quasi-periodic systems. On the other hand, in classical electric systems, dual and mixed transmission lines have been studied [78,87] using a Fibonacci distribution of two different values of inductances L_A and L_B, namely

$$L_A L_B L_A L_A L_B L_A L_B L_A L_A L_B L_A L_A L_B \ldots \tag{57}$$

Notice that when we introduce disorder in the inductances of the dual or mixed TL, the generic Equation (3) shows that the disorder appears simultaneously in the diagonal and non-diagonal part.

In dual transmission lines, the localization behavior of the Fibonacci quasi-periodic distribution of inductances L_n, keeping constant the capacitances $C_n = C_0 \; \forall n$, has been studied [78] analyzing the spectrum of the generalized Rényi entropies $R_q(\omega)$ versus ω and the spectrum of the inverse participation ratio $N \times IPR(\omega)$ versus ω. For each q value, $R_q(\omega)$ and $N \times IPR(\omega)$ show more than four global sub-bands. This happens because the allowed frequency band of the dual transmission lines is unbounded from above, namely every frequency of the spectrum is greater than a critical frequency ω_c, i.e., $\omega > \omega_c$. At the same time, the spectrum of $R_q(\omega)$ and $N \times IPR(\omega)$ clearly shows the self-replication behavior, where each sub-band is divided into three sub-bands until it is resolvable (see Figures 2 and 3 of Ref. [78]). This localization behavior is characteristic of quasi-periodic Fibonacci systems. Inside each sub-band, we find extended and localized states and gaps. When system size N grows, the number of gaps and localized states increases in such a way that the integrated density of states $IDOS(\omega)$ behaves in a fractal way. As a consequence, the total bandwidth goes to zero in the thermodynamic limit $N \to \infty$. In Figure 3a) we show the Shannon entropy $S(\omega)$ (39), which corresponds to the $R_1(\omega)$ Rényi entropy discussed in Ref. [78]. Notice that the number of global sub-bands is greater than four. Figure 3b) shows the three sub-bands existing in the global sub-band indicated by the vertical arrow in Figure 3a). These previous results about the number of global sub-bands of the dual TL shown in Ref. [78] change when the Fibonacci disorder is introduced in the mixed transmission line [87]. However, the self-replication of the spectrum is maintained for all kinds of mixed TL formed by p direct cells and q dual cells.

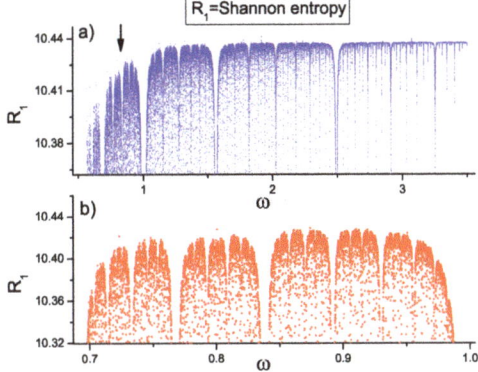

Figure 3. (a) Global sub-band structure of the Shannon entropy $S(\omega) = R_1(\omega)$ versus ω, for the Fibonacci quasi-periodic distribution of inductances L_n discussed in Ref. [78]. (b) Self-replication structure of the sub-band indicated by the vertical arrow in (a).

Remember that mixed TL are generated by a repetition pattern formed by a group of successive p direct cells, followed by a group of successive q dual cells. This topology generates a spectrum of allowed frequencies formed by exactly $d = (p+q)$ bands, as indicated in Section 2.4. In Ref. [87], only the q inductances of the dual cells of mixed TL were distributed according to the Fibonacci sequence, keeping constant the values of the other capacitances and inductances. The localization behavior of the average overlap amplitude (NC_ω) versus ω for the case $p = 2$ and $q = 1$, shows three ($d = 3$) allowed bands (see Figure 10 of Ref. [87]). These d bands exist regardless of the type of disorder and the degree of correlation; however, the position of these d bands depends on the values of all capacitances ($C_{n,x}, C_{n,y}$) and inductances ($L_{n,x}, L_{n,y}$) of direct cells (labeled x) and dual cells (labeled y), and the values of p and q. The average overlap amplitude (NC_ω) versus ω shows four global sub-bands, where each sub-band is divided into three sub-bands until it is resolvable (see Figure 11 of Ref. [87]). This result coincides with the one obtained from the quantum tight-binding model with diagonal Fibonacci disorder. This coincidence occurs because both models (mixed TL and tight-binding model) have bounded spectra and because in both models the Fibonacci disorder appears in the diagonal part of the corresponding dynamic equations (Equations (3) and (5)). Conversely, this result is different from the case of the dual transmission line shown in Figures 2 and 3 of Ref. [78], where the number of global sub-bands is greater than four, because the frequency spectrum of the dual TL is unbounded from the above. Interestingly, when p and q change, each of the $d = (p+q)$ bands of the mixed TL can accommodate a different number of global sub-bands. However, the self-replication is always present. To observe this behavior, let us consider the case with fixed $p = 2$, and two different values of q, namely $q = \{2, 3\}$. Figure 4 shows the average overlap amplitude (NC_ω) versus ω, for (a) $q = 2$ ($d = 4$ bands) and (b) $q = 3$ ($d = 5$ bands). There we can see that the full spectrum of frequencies of mixed TL is contained only within the d bands.

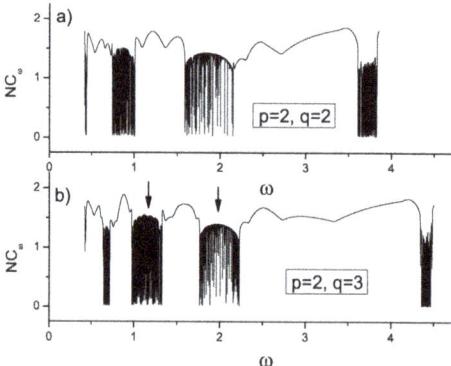

Figure 4. (NC_ω) versus ω for the Fibonacci distribution of inductances L_n, for mixed TL with fixed $p = 2$, considering two values of q. (a) $q = 2$ ($d = 4$ bands) and (b) $q = 3$ ($d = 5$ bands). Vertical arrows indicate the bands to be studied in Figure 5.

Figure 5 shows (NC_ω) versus ω for the third band and fourth bands shown in Figure 4b. In addition, Figure 5b,d show the self-replication behavior of each sub-band indicated with a vertical arrow in Figure 5a,c, respectively. In this figure we can see that the number of localized states and gaps increases after each self-replication.

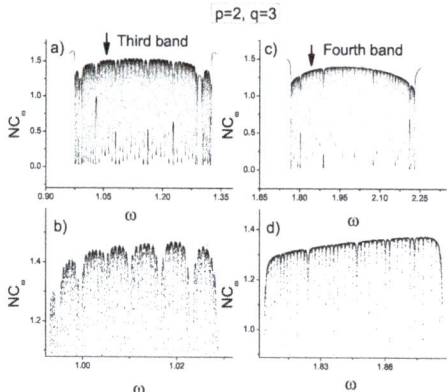

Figure 5. (NC_ω) versus ω for the Fibonacci distribution of inductances L_n for mixed TL, for $p = 2, q = 3$. A detail of Figure 4. (a) Third sub-band, (c) fourth sub-band. Figures (b,d) show the self-replication of the sub-bands indicated by vertical arrows in (a,c), respectively.

In summary, for arbitrary values of p and q forming the mixed TL, the set of $d = (p + q)$ bands accommodates the full spectrum generated by the Fibonacci distribution of inductances L_n. Inside each of the d bands, the number of global sub-bands is always greater than or equal to four, and the self-replication behavior corresponding to quasi-periodic systems is always present. In the self-replication process, new localized states and gaps appear repeatedly. Consequently, the integrated density of states has a fractal behavior and $IDOS(\omega) \to 0$ in the thermodynamic limit.

3.1.2. Generalized Thue–Morse Sequence

The generalized Thue–Morse (TM) aperiodic sequence can be generated by means of the substitution rule $A \to A^m B^n$, $B \to B^n A^m$. The corresponding substitution M matrix is given by

$$M = \begin{bmatrix} m & n \\ m & n \end{bmatrix}$$

The λ maximal eigenvalue of M is $\lambda = (m+n)$. For the case $m = 1$ and $n = 1$, we obtain the usual Thue–Morse sequence: $A \to AB$, $B \to BA$, namely

$$A \to AB \to ABBA \to ABBABAAB \to \cdots$$

The maximal eigenvalue $\lambda = (m+n) = 2$ is thus the length of the substitution, which means that $N = \lambda^k$ is the number of the A and B letters in the kth iteration. For the generalized Thue–Morse sequence, the relative frequency of both types of letters, n_A and n_B, is the following $n_A = \frac{m}{(m+n)}$ and $n_B = \frac{n}{(m+n)}$. Another generalization of the Thue–Morse sequence is the m−tupling sequence generated by the substitution rule $A \to AB^{m-1}$, $B \to BA^{m-1}$, with $m \geq 2$. In this case, the maximal eigenvalue of the corresponding substitution M matrix is $\lambda = m$, and the number of letters N in this sequence also increases geometrically, i.e., $N = m^k$, where k is the iteration order. Here $n_A = n_B = \frac{1}{2}$. For $m = 2$ we return to the usual Thue–Morse sequence. A spectral classification of one-dimensional binary aperiodic crystals as a function of the substitution matrix M is shown in Ref. [73].

For the tight-binding quantum model, the aperiodic properties of the generalized Thue–Morse systems have been studied in great detail by [47–58,60,62,71]. Additionally, in classical dual and direct transmission lines the Thue–Morse and the m−tupling distribution of capacitances and inductances have been studied by [83–85]. For direct TL, two values of inductances L_A and L_B, where distributed according to the m−tupling substitution rule $L_A \to L_A L_B^{m-1}$, $L_B \to L_B L_A^{m-1}$, $m \geq 2$, keeping constant the capacitances [83,84]. For $m = 2$ we obtain the usual Thue–Morse substitution rule $L_A \to L_A L_B$, $L_B \to L_B L_A$. One of the principal findings of these studies was that the localization properties of the usual Thue–Morse case, namely $m = 2$, is markedly different to the $m = 3$ case. In general, although in the m−tupling sequence the number of letters A and B in each iteration is the same ($n_A = n_B$) for any value of m, the number of extended states in the m−tupling inductance distribution depends on the specific value of m. This was demonstrated numerically using different localization tools, like normalized localization length $\Lambda(\omega)$, participation number $D(\omega)$, normalized participation number $\xi(\omega)$, global density of states $DOS(\omega)$, transmission coefficient $T(\omega)$ and the average overlap amplitude (NC_ω). In addition, it was shown that inside the m−tupling family, starting with $m = 3$, the number of extended states increases as the value of m increases, so that for $m \gg 3$, the allowed spectrum is similar to the spectrum of the case $m = 2$ (Thue–Morse). This can be seen in Figure 6 where we show the normalized participation number $\xi(\omega)$ for three values of m, namely $m = \{2,3,13\}$ (a) to (c). Also, in Figure 6d we indicate with a short vertical bar the spectrum of the extended states, namely the frequencies for which the $\Lambda(\omega)$ normalized localization length meets the condition $\Lambda(\omega) \geq 1$. The image shows that the number of extended states for $m = 3$ is small compared to the case $m = 2$. However, for the case $m \gg 3$, namely $m = 8$ and $m = 13$, the number of extended states becomes comparable with case $m = 2$.

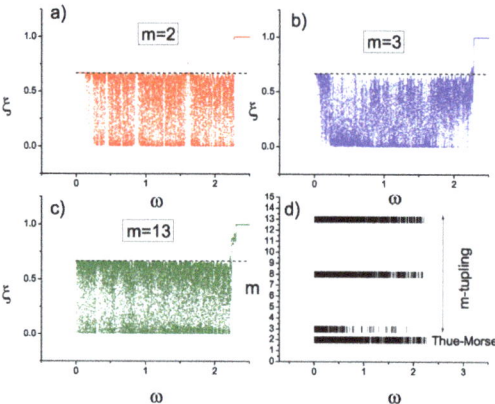

Figure 6. $\xi(\omega)$ versus ω for the m−tupling distribution of inductances L_n in the direct TL, for three values of m, namely $m = \{2, 3, 13\}$ (**a**–**c**). (**d**) $\Lambda(\omega)$ versus ω. A short vertical bar indicates the existence of an extended state ($\Lambda(\omega) \geq 1$). The number of extended states for $m = 3$ is very small compared to the case $m = 2$. Conversely, for $m \gg 3$ ($m = 8$ and $m = 13$), the number of extended states increases and becomes comparable to the $m = 2$ case.

When comparing the spectrum for cases $m = 2$ and $m = 13$, in a restricted region of frequencies (see Figure 7), it can be observed that the number of extended states which fulfills the condition $\xi \approx 0.667$, corresponding to the periodic case, is reasonably similar in both cases. Also, we can see that the sub-band of extended states of the Thue–Morse case with $m = 2$ (Figure 7a) is much wider than the sub-bands of extended states of the m−tupling case with $m = 13$ (Figure 7b).

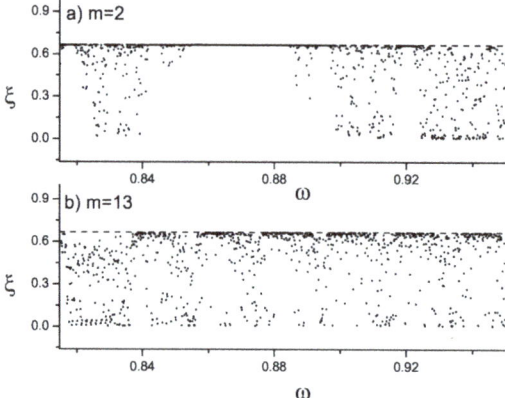

Figure 7. $\xi(\omega)$ versus ω for m−tupling distribution of inductances L_n in direct TL in a restricted region of frequencies of the Figure 6. (**a**) $m = 2$, (**b**) $m = 13$. We can see that the sub-bands of extended states ($\xi \approx 0.667$) for $m = 2$ is much wider than the sub-bands of extended states of case $m = 13$.

In sum, for direct transmission lines with m−tupling distribution of inductances, the frequency spectrum of the Thue–Morse sequence ($m = 2$) can be considered the limit of the m−tupling sequence's frequency spectrum when $m \gg 3$. On the other hand, the number of extended states for the case $m = 2$ decreases dramatically when m changes to $m = 3$, as shown in Refs. [83,84] and Figure 6.

As an extension of these ideas, the localization behavior of dual transmission lines with non-linear capacitances has been studied [85]. The non-linear behavior of capacitances is introduced through the $V_{C,n}$ potential difference across each capacitance, i.e.,

$$V_{C,n} = q_n \left(\frac{1}{C_n} - \varepsilon_n |q_n|^2 \right) \quad (58)$$

C_n is the linear part of the capacitance $V_{C,n}$ and ε_n is the amplitude of the non-linear term. The equation corresponding to this dual case is given by

$$\left(L_{n-1} + L_n - \frac{1}{\omega^2 C_n} + \frac{\varepsilon_n |I_n|^2}{\omega^4} \right) I_n - L_{n-1} I_{n-1} - L_n I_{n+1} = 0 \quad (59)$$

When the non-linear amplitudes ε_n go to zero ($\varepsilon_n \to 0$), we return to the dual linear Equation (2). The localization behavior of this non-linear dual TL has been studied using two values of the non-linear amplitude ε_n, namely ε_A and ε_B, distributed according to the m−tupling Thue–Morse sequence [85], i.e., $\varepsilon_A \to \varepsilon_A \varepsilon_B^{m-1}$, $\varepsilon_B \to \varepsilon_B \varepsilon_A^{m-1}$, $m \geq 2$, but keeping constant the capacitances C_n and inductances L_n, namely $C_n = C_0$, and $L_n = L_0$ $\forall n$. In this case, the aperiodic disorder appears only in the diagonal term of the dynamic Equation (59).

The same fundamental result about the localization degree of the $m = 2$ case in comparison with the $m \geq 3$ case reappears in this non-linear case, that is, for fixed values of ε_A and ε_B, the $m \geq 3$ family does not belong to the family corresponding to $m = 2$, and in addition, for $m \gg 3$ the frequency spectrum begins to resemble the spectrum of the case $m = 2$. To be specific, for $m = 2$ we can see a large number of extended states across the entire frequency spectrum, mixed with localized states and gaps. On the contrary, for $m = 3$, almost the entire frequency spectrum is formed with localized states and gaps, accordingly showing a huge decrease in the number of extended states. This behavior can be seen in Figure 6 of Ref. [85] that shows $\xi(\omega)$ versus ω for $m = \{2, 3, 5, 9\}$. To compare the cases for $m = 2$ and $m = 3$ in more detail, in Figure 8 we show the average overlap amplitude (NC_ω), for $m = \{2, 3\}$ with $\varepsilon_A = 0.1$ and $\varepsilon_A = 0.03$, keeping constant the values of capacitances C_n and inductances L_n. When m changes from $m = 2$ to $m = 3$, the number of extended states decreases markedly, almost tending to zero. The horizontal dashed line corresponds to the periodic linear case, $\varepsilon_A = \varepsilon_B = 0$, for which (NC_ω) fulfills the condition $(NC_\omega) \approx 1.62$. Moreover, at the top of each figure, we indicate with a short vertical bar the presence of an extended state, namely $\Lambda(\omega) \geq 1$. Both results about the number and position of the extended states in each case match each other. This localization behavior coincides with the results shown in Figure 6 of Ref. [85], when studying the localization behavior of the normalized participation number $\xi(\omega)$ and $\Lambda(\omega)$. This way, we have demonstrated that the $m = 2$ (Thue–Morse) case is different than the $m = 3$ case (m−tupling).

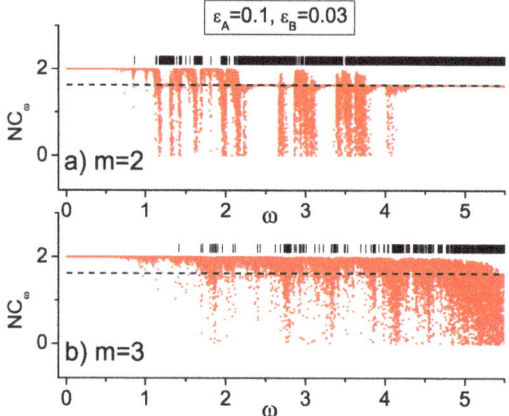

Figure 8. (NC_ω) versus ω for the m-tupling distribution of amplitudes ε_n of non-linear capacitances of the dual TL with $\varepsilon_A = 0.1$ and $\varepsilon_B = 0.03$. (**a**) Case $m = 2$ and (**b**) case $m = 3$. The horizontal dashed lines indicate the value (NC_ω) ≈ 1.62, for the periodic linear case, i.e., $\varepsilon_A = \varepsilon_B = 0.0$. In addition, at the top of each figure, we indicate with a vertical bar the presence of extended states which meet condition $\Lambda(\omega) \geq 1$.

Consider now the localization behavior of the integrated density of states $IDOS(\omega)$ for the non-linear case, with $\varepsilon_A = 0.1$ and $\varepsilon_A = 0.03$. In Figure 9a we show the $IDOS(\omega)$ for four values of m, namely $m = \{2, 3, 8, 13\}$. For each m we use a constant size $N_m = m^k$, namely $N_m = \{2^{21}, 3^{13}, 8^7, 13^6\}$. There we can see that for the case $m = 2$, the $IDOS(\omega)$ is always greater than the $IDOS(\omega)$ of any other value of $m \geq 3$. However, when m grows ($m = 8$ and $m = 13$), the $IDOS(\omega)$ grows, approaching the values for the case $m = 2$. This behavior confirms the conjecture that the Thue–Morse sequence ($m = 2$) can be considered to be a limit case of the m-tupling sequence when $m \gg 3$. We now turn to the behavior of the $IDOS(\omega)$ for fixed $m = 3$, as a function of the $N = 3^k$ system size, with $k = \{10, 12, 13\}$ (see Figure 9b). For the minimum value $k = 10$, the $IDOS(\omega)$ is the greatest of all, but when N increases (the value of k increases), the number of extended states decreases (the $IDOS(\omega)$ decreases), and new localized states and gaps appear that barely contribute to the integrated density of states. As a consequence, the $IDOS(\omega)$ tends to zero. This behavior is characteristic of aperiodic systems.

On the other hand, for fixed value of m, when the difference $|\varepsilon_A - \varepsilon_B|$ between the values of the amplitudes of the non-linear term increases, so does the disorder degree of the transmission line, which tends to localize the electric current function $I(\omega)$, and as a result, the integrated density of states $IDOS(\omega)$ go to zero. This behavior can be observed in Figure 5 of Ref. [85], for $m = 2$ (Thue–Morse case), considering a fixed value $\varepsilon_A = 0.0$ (the periodic linear case) and three different values of ε_B, namely $\varepsilon_B = \{0.01, 0.03, 0.07\}$.

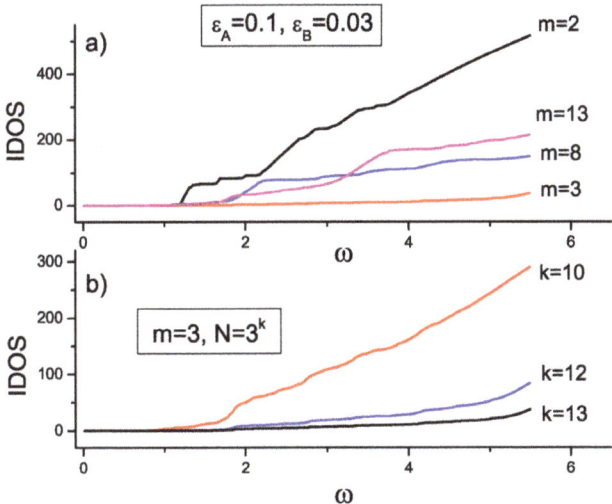

Figure 9. The integrated density of states $IDOS(\omega)$ versus ω for the m−tupling distribution of amplitudes ε_n of the non-linear capacitances of the dual TL for $\varepsilon_A = 0.1$ and $\varepsilon_B = 0.03$. (**a**) *IDOS* for four m values, i.e., $m = \{2, 3, 8, 13\}$. For each m, we used a fixed N_m, i.e., $N_m = \{2^{21}, 3^{13}, 8^7, 13^6\}$. The *IDOS* for $m = 2$ is the greatest of all, and the *IDOS* for $m = 3$ is the smallest of all. For increasing values of m ($m = 8$ y $m = 13$), the *IDOS* tends to the value corresponding to $m = 2$. (**b**) Fixed $m = 3$, as a function of the $N = m^k$, with $k = \{10, 12, 13\}$. The *IDOS* corresponding to $N = 3^{10}$ is the largest of all. When N increases, the *IDOS* tends to zero, $IDOS \to 0$.

3.1.3. Incommensurate Sequences

The aperiodic incommensurate systems are generated by two superimposed periodic structures with incommensurate periods. The origin of incommensurability may be structural or dynamic. In the first case, two or more superimposed periodic structures with incommensurate periods exist, and in the second case one periodicity is related to the crystalline structure and the other to the behavior of elementary excitations that propagate through the crystal. Two of the most studied incommensurate models are the Aubry–André model and the Soukoulis–Economou model.

In the one-dimensional tight-binding quantum model, the site energies ε_n have been distributed according to the Aubry–André model that is

$$\varepsilon_n = \varepsilon_0 + b \cos(2\pi\beta n) \tag{60}$$

where ε_0 is the single-site energy of the unperturbed periodic lattice, b is the amplitude and β is an irrational number, usually $\beta = 0.5\left(\sqrt{5} - 1\right)$ (the inverse of the Fibonacci golden mean). For $b = 2.0$, a phase transition from extended to localized states appears [36,69,70,72].

In classical electric transmission lines, the Aubry–André model has been used to distribute the inductances L_n in two different cases: (a) direct TL with constant capacitances $C_n = C_0\ \forall n$ (diagonal disorder) [86] and (b) mixed transmission lines with disorder only in the q inductances of the dual cells, keeping constant the value of all the other electrical components of the direct and dual cells [87]. In this case, the disorder appears in the diagonal and the off-diagonal terms of the generic Equation (3).

In case a), the inductances L_n of the direct transmission line are distributed according to the Aubry–André sequence:

$$L_n = L_0 + b\cos(2\pi\beta n) \tag{61a}$$

where $L_0 = const.$ and $b < L_0$. In this case, the aperiodic incommensurate disorder only appears in the diagonal term of the generic Equation (3). The localization behavior of this classic electric model can be visualized in Figure 10 where the map (b,ω) is shown for $L_0 = 4.0$ and $N = 7 \times 10^5$. Each dot on the map indicates the existence of an extended state, because the normalized localization length fulfills condition $\Lambda(\omega) \geq 1.0$. For $b \to 0$, the frequency spectrum shows a single band of extended states that corresponds to the periodic case. For increasing values of b, i.e., for $b \leq 1.9$, the map (b,ω) shows three global sub-bands of extended states (with localized states and gaps) separated by two large gaps. After that, for $b > 1.9$, only two global sub-bands of extended states survive, which also contain localized states and gaps. Finally, for b close to $b = L_0$, there is only a small sub-band where almost all states are extended states, namely $0 \leq \omega \leq 0.45$.

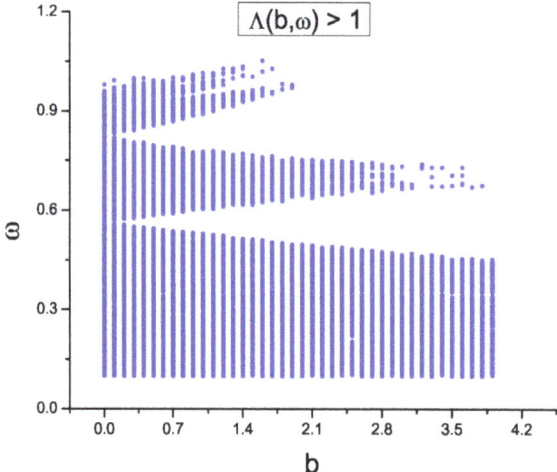

Figure 10. Map (b,ω) for the Aubry–André distribution of inductances with $L_0 = 4.0$. Each point of the map indicates an extended state, because $\Lambda(\omega) \geq 1.0$. For increasing values of the amplitude b, the number of sub-bands of extended states diminishes, and for b close to $b = L_0$, there is only a small sub-band where almost all states are extended states, namely $0 \leq \omega \leq 0.45$.

Figure 11 shows (a) the $\lambda(\omega)$ Lyapunov exponent versus ω for two values of the b amplitude $b = \{1.5, 3.99\}$ and b the spectrum of the extended states, $\Lambda(\omega) \geq 1$ versus ω for fixed $b = 1.5$. Figure 11a shows that for $b = 3.99 \approx L_0$ (thick red line), only one band of extended states ($\lambda(\omega) \to 0$) can be observed for $\omega \leq 0.45$. Conversely, for $\omega > 0.45$, only gaps and localized states can be found for this value of b. This result coincides with the result indicated by the map shown in Figure 10. On the other hand, in the same Figure 11a we draw $\lambda(\omega)$ versus ω for a smaller value of b, namely $b = 1.5$ (thin black line). There we can see several sub-bands of extended states ($\lambda(\omega) \to 0$) separated by gaps. Within these sub-bands, we can find more localized states and gaps, which are not perceived in this picture. On the contrary, these gaps can be seen in Figure 11b, where a detail of the map Figure 10 is shown for $b = 1.5$, for fixed $N = 10^6$. In this figure, each short vertical bar indicates an extended state, because $\Lambda(\omega) \geq 1.0$. The vertical dashed arrows that cross both figures (for the case $b = 1.5$) indicate the edge of the gaps, i.e., the frequencies for which phase transitions occur.

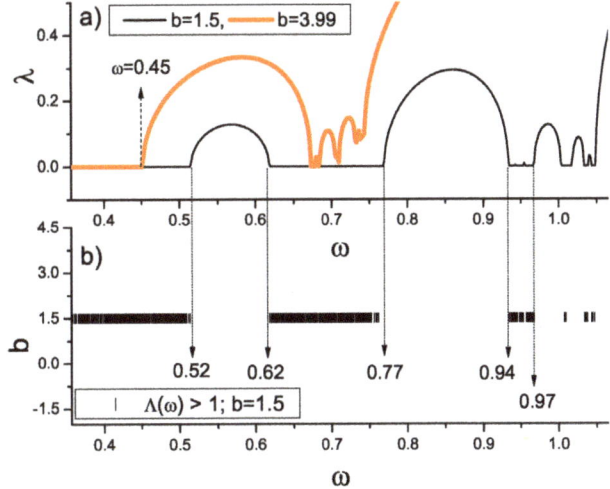

Figure 11. (**a**) Lyapunov exponent $\lambda(\omega)$ versus ω for two values of the b amplitude $b = \{1.5, 3.99\}$. For $b = 3.99 \approx L_0$ (thick red line), only one band of extended states ($\lambda(\omega) \to 0$) can be observed for $\omega \leq 0.45$. Conversely, for $b = 1.5$ (thin black line), we can see several sub-bands of extended states ($\lambda(\omega) \to 0$) separated by gaps. Within these sub-bands, we can find more localized states and gaps. (**b**) The spectrum of the extended states, $\Lambda(\omega) \geq 1$ versus ω for fixed $b = 1.5$ and $N = 10^6$. The vertical dashed arrows that cross both images indicate the edge of the gaps, in which phase transitions occur.

To see in more detail the phase transition from extended to localized states, in Figure 12a we show the Lyapunov exponent $\lambda(\omega)$ versus ω for the cases $b = 2.0$ and $N = 2 \times 10^5$. The vertical arrows indicate the frequencies $\omega_1 = 0.5006231$, $\omega_2 = 0.6336884$, and $\omega_3 = 0.7577136$, to be studied in Figure 12b–d, respectively. In these last three images, we show the scaling behavior of the average overlap amplitude (NC_ω) for three values of N, namely $N = \{8, 12, 16\} \times 10^4$. For each frequency ω_1, ω_2 and ω_3, we find a phase transition from extended states to localized states at the critical value $b_c = 2.0$. To the left of the critical point b_c for almost every amplitude b, all (NC_ω) values coalesce into a single one, i.e., $(NC_\omega) \to const > 0$, indicating an extended behavior. On the contrary, to the right of the critical point ($b > b_c$), (NC_ω) grows as system size N grows, indicating a localized behavior.

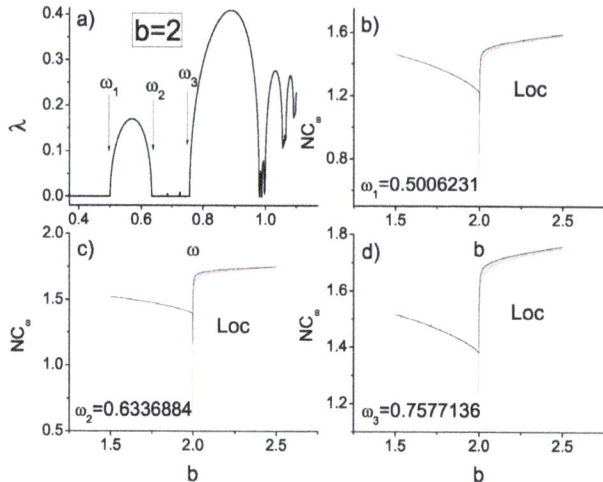

Figure 12. (a) Lyapunov exponent $\lambda(\omega)$ versus ω for the case $b = 2.0$ and $N = 2 \times 10^5$. The vertical arrows indicate the frequencies $\omega_1 = 0.5006231$, $\omega_2 = 0.6336884$, and $\omega_3 = 0.7577136$, to be studied in (b–d). In these last three images, we show the scaling behavior of (NC_ω) for three values of N, namely $N = \{8, 12, 16\} \times 10^4$. For each frequency ω_1, ω_2 and ω_3, we can see a phase transition from extended states to localized states at the critical value $b_c = 2.0$. For $b \leq b_c$ we find only extended states, because for almost every amplitude b, all (NC_ω) values coalesce into a single one, i.e., $(NC_\omega) \to const. > 0$. On the contrary, for $b > b_c$, (NC_ω) grows as system size N grows, indicating a localized behavior.

These results coincide with those in Figures 4–6 of Ref. [86], where this same problem was studied. In [86], a phase transition from the extended to the localized state is found depending on amplitude parameter b. This result was found for different frequency values, by studying transmission coefficient $T(\omega)$ and the scaling behavior of the average overlap amplitude (NC_ω).

In case (b), for mixed transmission lines (with p direct cells and q dual cells), the inductances L_n of the q dual cells were distributed according to the Aubry–André sequence [87], namely $L_{n,y} = L_{0,y} + b \cos\cos(2\pi\beta n)$ with $b < L_{0,y}$. All other electric components are kept constant, i.e., for direct cells $L_{n,x} = L_{0,x}$, $C_{n,x} = C_{0,x}$ and for dual cells $C_{n,y} = C_{0,y}$. In Ref. [87], three different cases were studied: (a) $p = 2$, $q = 1$, (b) $p = 1$, $q = 4$ and (c) $p = 4$, $q = 1$. In all cases, the frequency spectrum is completely contained within the $d = (p + q)$ bands generated by the mixed TL. For fixed b, in each of the d bands, it is always possible to find sub-bands of extended states in addition to localized states and gaps. These results were obtained by studying the transmission coefficient $T(\omega)$ and the scaling of the average overlap amplitude (NC_ω) (see Figures 7–9 of Ref. [87]). To see the influence of the b amplitude in the localization behavior, in Figure 13 we show the transmission coefficient $T(\omega)$ for four values of b, namely $b = \{0.3, 0.7, 1.1, 1.5\}$ for the case $p = 2$, $q = 1$. We use the same values of the electric components used in Ref. [87]. In particular, $L_{0,y} = 1.6$. For $b = 0.3$ we find $d = 3$ bands containing extended states, localized states and gaps (similar to Figure 7a of Ref. [87]). However, for increasing values of b, the number of extended states within each band decreases, and as a consequence both lateral bands begin to disappear. In this way, for $b = 1.1$, the leftmost band has already disappeared, and for $b = 1.5$, the rightmost band is about to disappear.

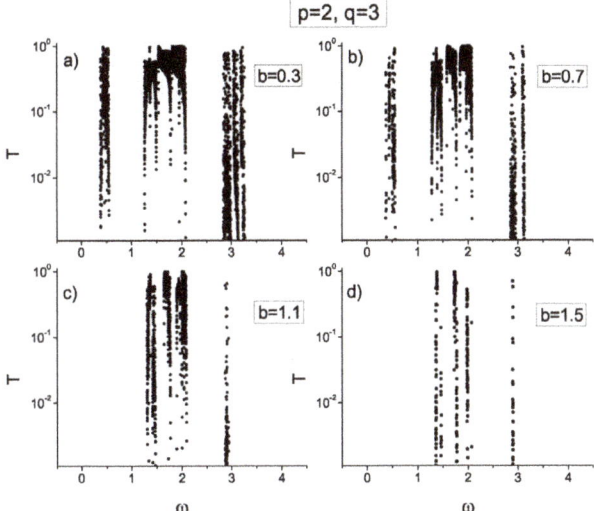

Figure 13. Transmission coefficient $T(\omega)$ versus ω, for mixed transmission line with $p = 2$, $q = 1$, for (**a**) $b = 0.3$, (**b**) $b = 0.7$, (**c**) $b = 1.1$ and (**d**) $b = 1.5$. For (**a**) $b = 0.3$ we find $d = (p+q) = 3$ bands containing extended states, localized states and gaps. For increasing values of b, the number of extended states within each band decreases. Therefore, for (**c**) $b = 1.1$, the leftmost band has already disappeared, and for (**d**) $b = 1.5$, the rightmost band is about to disappear.

3.2. Long-Range Correlated Disorder

For one-dimensional disordered systems without any correlation in the disorder (white noise), all states are localized states in the thermodynamic limit. However, the introduction of correlation in the disorder can trigger the appearance of a discrete set of extended states (short-range correlation) or bands of extended states (long-range correlation). The correlated disorder has been introduced in quantum tight-binding systems [2–29] and in classical systems such as harmonic chains [96–99], and electrical transmission lines [76,77,79,80,88].

The quantum tight-binding Equation (5) and the generic Equation (3) describing transmission lines are similar. Transformations (6) and (7) permit the correspondence between both models. However, unlike the quantum case, in transmission lines it is impossible to study the pure off-diagonal case, because the disorder contained in the vertical impedances (the coupling between neighboring electric cells) appears in the off-diagonal coefficients B_{n-1} and B_n of the generic Equation (3) and in the diagonal coefficient $D_n = (B_{n-1} + B_n - A_n)$ too.

To analyze the main differences in the localization behavior with the one-dimensional quantum case, the dual, direct and mixed disordered transmission lines have been studied recently. These studies include long-range correlated disorder and diluted disordered TL. In addition to continuous sequences, the long-range correlation has been used to generate discrete sequences (binary and ternary).

3.2.1. Discrete Sequences

To generate long-range correlated sequences $\{x_n\}$ we use the Fourier filtering method (FFM). Let us consider initially a set of uncorrelated random numbers $\{u_n\}$ with a Gaussian distribution. Then we take the fast Fourier transform (FFT) of the random sequence $\{u_n\}$ and we obtain a new sequence $\{u_k\}$. The long-range correlation is introduced in the sequence $\{u_k\}$ doing the following transformation $x_k = u_k k^{-\frac{1}{2}(2\alpha-1)}$. Calculating the inverse FFT of the new sequence $\{x_k\}$, we obtain the long-range correlated sequence $\{x_n\}$ which is spatially correlated with the $S(k)$ spectral density

$S(k) \propto k^{-(2\alpha-1)}$. Here the exponent α of the transformation is known as the correlation exponent and fulfills the condition $\alpha \geq 0.5$. For $\alpha = 0.5$, we regain the uncorrelated random sequence (white noise). Correlation exponent α quantifies the degree of long-range correlation imposed in the original random sequence $\{u_n\}$. Finally, we normalize the correlated sequence $\{x_n\}$ to obtain zero average, $\langle x_n \rangle = 0$, and the variance is set to unity.

From the long-range correlated sequence $\{x_n (\alpha > 0.5)\}$, we can generate the asymmetric ternary sequence $\{v_n (b_1, b_2)\}$ formed with three letters, A, B and C,

$$v_n = \begin{cases} A & \text{if } x_n < b_1 \\ C & \text{if } b_1 \leq x_n \leq b_2 \\ B & \text{if } x_n > b_2 \end{cases} \tag{62}$$

with $b_2 \geq b_1$. The symmetric ternary map is obtained when $b_2 = -b_1 = b$. If $b_1 = b_2 = b$ we obtain the asymmetric binary sequence $\{A, B\}$. For $b \to 0$ we regain the symmetric binary sequence. Please note that the long-range correlation of the ternary sequence $\{v_n (b_1, b_2)\}$ is not exactly quantified by the correlation exponent α, because the map (62) changes the long-range correlation. In one-dimensional tight-binding systems, the symmetric binary and ternary model has been studied [8,24,25,28]. In these models, a metal-insulator transition has been reported as a function of the correlation degree α and size b of the window. In addition, the asymmetric ternary map (62) was studied using electrical dual transmission lines [76] considering three values of capacitances $C_n = \{C_A, C_B, C_C\}$, maintaining constant the inductances $L_n = L_0$ $\forall n$. This case contains only diagonal disorder. The long-range correlation in the distribution of capacitances was generated through the FFM. For TL with a finite number of cells, it is possible to find bands of extended states whose size increases for increasing values of correlation exponent α. For the asymmetrical model, the normalized localization length $\Lambda (\omega, \alpha, b_1, b_2, N)$ is a complicated function of the parameters ω, α, b_1, b_2, and N, but for fixed frequency ω, for $N \to \infty$, it is always possible to find a transition from localized electric current functions to extended current functions for some specific values of the parameters. For the symmetrical ternary map $b_2 = -b_1 = b$, a phase diagram (α, b) separating localized states from extended states has been found for fixed frequency performing finite-size scaling of the normalized localization length $\Lambda (\omega)$. This result is similar to the phase diagram found in the tight-binding case.

Moreover, the same ternary dual TL was studied, but using the Ornstein–Uhlenbeck method to generate the long-range correlation [77]. In this method, the degree of long-range correlation depends on two independent parameters, i.e., the viscosity coefficient γ and the diffusion coefficient C. Studying the scaling behavior of $\Lambda (\omega)$, we obtain two-phase diagrams for the symmetrical map when C and γ are independent parameters, namely (C, b) for fixed γ and (γ, b) for fixed C. In addition, we study the phase diagrams when C and γ are dependent parameters, i.e., $C = \gamma^2$. In all cases, we find a transition from localized to extended states. Also, the harmonic symmetric ternary chain was studied in Ref. [99] using the Ornstein–Uhlenbeck method for the case $C = \gamma^2$. Instead of the transition from localized to extended behavior, they found a disorder-order transition for $b > 4$, because the disorder degree practically disappears at this limit.

This same kind of disorder-order transition has been found by studying localization properties of direct TL with diluted and non-diluted asymmetric dichotomous noise (binary sequences of inductances L_A and L_B with $C_n = C_0$ $\forall n$) [82]. The asymmetric dichotomous sequence $\{\zeta (t)\}$ is generated by a variable $\zeta (t)$ which switches in time in a random way between two given values a and $(-b)$ with transition rates μ_a and μ_b, respectively (dichotomous noise). Considering $\zeta (t)$ as a stationary process, the dichotomous noise has zero mean and is exponentially correlated. The τ correlation time of the dichotomous noise is defined as $\tau^{-1} = (\mu_a + \mu_b)$. In addition, from the zero-average condition $\langle \zeta (t) \rangle = 0$, we obtain the following relationship between a, b, μ_a and μ_b, namely $\beta = \frac{a}{b} = \frac{\mu_a}{\mu_b}$, where parameter β measures the degree of asymmetry of the dichotomous noise. For $\mu_a > \mu_b$ we have $\beta > 1$ and for $\mu_a < \mu_b$ we have $\beta < 1$. The symmetric sequence $\mu_a = \mu_b$ is obtained for the case $\beta = 1$. Consequently, the dichotomous noise is characterized by three independent parameters: τ,

a and b. However, setting the value of one of the parameters, for example, $b = 1$, we can study the localization behavior generated by this kind of exponentially correlated noise using only two independent parameters, i.e., τ and β. In the diagonal disordered direct TL, the inductances L_A and L_B are distributed according to the asymmetric dichotomous noise, keeping the capacitances constant $C_n = C_0 \, \forall n$ [82]. For $\tau < \tau_c$ and for $\beta < \beta_c$ (τ_c, β_c are critical values) the electric current function $I(\omega)$ shows a localized behavior, but for $\tau > \tau_c$ and for $\beta > \beta_c$, the $D(\omega, \tau, \beta)$ participation number scales as $D(\omega) \propto N^{m(\omega, \tau, \beta)}$, where $m(\omega, \tau, \beta) < 1$ is the slope of the linear relationship between $\ln(D(\omega, \tau, \beta))$ and $\ln(N)$, for fixed ω, β, and τ. Only in the limit $\tau \to \infty$ (for fixed ω and β) and $\beta \to \infty$ (for fixed ω and τ), we obtain the exact linear behavior, i.e., $\lim_{\tau \to \infty} m(\omega, \tau, \beta) = 1.0$ and $\lim_{\beta \to \infty} m(\omega, \tau, \beta) = 1.0$. However, in both limits, $\tau \to \infty$ or $\beta \to \infty$, the asymmetric dichotomous sequence becomes a periodic sequence. Thus, we only can observe a disorder-order transition, which in turn indicates that all states are localized states in the thermodynamic limit for classical electric TL. This result coincides with the one obtained for the one-dimensional tight-binding quantum model with symmetric dichotomous noise, in which the metal-insulator transition is absent [56,100].

3.2.2. Continuous Sequences

In addition to discrete sequences, continuous long-range correlated sequences have been used to study the localization behavior of direct, dual and even mixed electrical transmission lines [79,80,88]. In general, in classical electric transmission lines, the long-range correlated disorder in capacitances and inductances has been used in the following form: $C_n(\alpha) = C_0 + b f(x_n(\alpha))$ and $L_n(\beta) = L_0 + b f(y_n(\beta))$, where $f(u)$ is an harmonic function. $\{x_n(\alpha)\}$ and $\{y_n(\beta)\}$ are two independent long-range correlated sequences generated by the FFM and α and β are the corresponding correlation exponents that determine the correlation degree. b is the amplitude of the fluctuation of C_n and L_n around C_0 and L_0, respectively. The diagonal and off-diagonal disordered dual transmission line, considering only one type of correlated sequence $\{x_n(\alpha)\}$ has been studied recently [79]. In this case, C_n and L_n vary in phase, i.e.,

$$C_n = C_0 + b \sin(2\pi x_n(\alpha)) \quad (63)$$
$$L_n = L_0 + b \sin(2\pi x_n(\alpha))$$

Here, $b < \min(C_0, L_0)$ to avoid negative values of the electrical components. For this kind of disorder, it is always possible to find extended states for different frequencies, and for each specific frequency a phase diagram (b, α), which separates extended states from localized states in the thermodynamic limit can be found.

To obtain the critical correlation exponent α_c separating localized states from extended states, we analyze the scaling behavior of (a) the participation number $D(\omega)$ (37), (b) the relative fluctuation $\eta_D(\omega, b, \alpha, N)$ of the participation number $D(\omega)$ and (c) the Binder cumulant $B_D(\omega, b, \alpha, N)$ of the participation number $D(\omega)$. These quantities are defined as

$$\eta_D(\omega, b, \alpha, N) = \sqrt{\left(\frac{\langle D^2 \rangle}{\langle D \rangle^2} - 1\right)} \quad (64)$$

and

$$B_D(\omega, b, \alpha, N) = \left(1 - \frac{\langle D^4 \rangle}{3 \langle D^2 \rangle^2}\right) \quad (65)$$

where $\langle .. \rangle$ means an average over long-range correlated sequences.

For increasing system size N, the relative fluctuation η_D goes to zero for extended states and grows converging to a finite value for localized states. Consequently, for $N \to \infty$, η_D tends toward a step function and a discontinuity appears that separates extended states from localized states. This scaling behavior can be used to determine the critical correlation exponent α_c for fixed values of ω

and b, because the curves $\eta_D(\alpha)$ with different N values will cross in a single point (the critical point α_c). Also, the scaling behavior of the Binder cumulant $B_D(\alpha)$ indicates that for $N \to \infty$, $B_D(\alpha)$ jumps abruptly from a constant value ($B_D = 0.667$) for extended states to zero ($B_D(\alpha) = 0$) for localized states. Consequently, for fixed values of ω and b, the curves $B_D(\alpha)$ with different N values will cross in a single point (critical point α_c). On the other hand, the critical value of the fluctuation amplitude b_c can be obtained studying the scaling behavior of the normalized localization length $\Lambda(b)$ (35). For fixed ω and $\alpha \geq \alpha_c$, in the transition point from localized to extended states, $\Lambda(b)$ varies from $\Lambda(b) > 1$ to $\Lambda(b) \to 0$. Finally, in the thermodynamic limit, for fixed frequency ω, the phase diagram (b, α) is formed by two independent straight lines, so that extended states only appear when condition $\alpha \geq \alpha_c$ is met for any $b \leq b_c$. Specifically, in Ref. [79] the following values were used: $C_0 = 0.5$, $L_0 = 1.0$. For the fixed frequency $\omega = 3.6$, the critical values are $b_c = 0.43$ and $\alpha_c = 1.51$ (see phase diagram in Figure 9 of Ref. [79]). In Figure 14 we show, in a schematic way, the phase diagram for a fixed frequency ω, when C_n and L_n vary in phase (63) in dual TL. This map is conceptually different to the map in Figure 15b), when C_n and L_n vary out of phase (in the study of mixed TL).

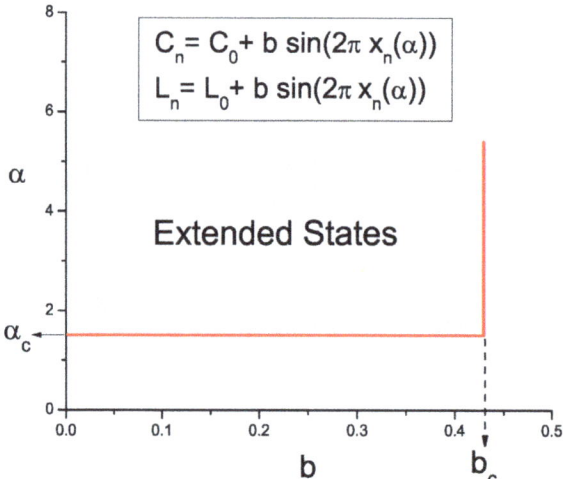

Figure 14. Schematic phase diagram (b, α), for a fixed frequency ω, when C_n and L_n vary in phase in dual transmission lines. For this transmission line with long-range correlated distribution of C_n and L_n, extended states can only be found for $\alpha \geq \alpha_c$ and $b \leq b_c$.

In Ref. [88], this model was generalized in two ways: (a) studying mixed transmission lines instead of dual TL, and (b) the capacitances $C_{n,y}$ and inductances $L_{n,y}$ of the dual cells of mixed TL are distributed out of phase, using two independent long-range correlated sequences $x_n(\alpha) \neq y_n(\beta)$. Specifically,

$$C_{n,y} = C_{0,y} + b \cos(2\pi x_n(\alpha)) \qquad (66)$$
$$L_{n,y} = L_{0,y} + b \cos(2\pi y_n(\beta))$$

where $\{x_n(\alpha)\}$ and $\{y_n(\beta)\}$ are two independent long-range correlated sequences, even in the case $\alpha = \beta$, because each correlated sequence is initiated using two independent uncorrelated random sequences according to the FFM. The localization behavior of this mixed TL was studied in Ref. [88] for the case $p = 2$, $q = 3$. The frequency spectrum of this case shows $d = (p+q) = 5$ bands. Additionally, in the thermodynamic limit, for fixed p, q and b, it is always possible to find an asymmetric phase diagram (α, β) for each frequency ω, corresponding to an extended state. In the case studied in Ref. [88], for $\omega = 1.986591$ and $b = 0.1$, the correlation exponents α and β fulfill the following condition: $\alpha \geq \alpha_c$ and $\beta \geq \beta_c$, with the asymmetric condition $\alpha_c \geq \beta_c$. This behavior can be observed in Figures 8 and 9 of Ref. [88]. There we can see the phase diagram (α, β) that separates localized states from extended states, and the localization behavior of $\Lambda(\omega)$ versus α (for fixed β), and $\Lambda(\omega)$ versus β (for fixed α). The asymmetric condition $\alpha_c \geq \beta_c$ can be explained through the following arguments. In relationships (66), the fluctuation $\Delta C_{n,y}$ of the capacitances around $C_{0,y}$ is the same as the fluctuation of inductances $\Delta L_{n,y}$ around $L_{0,y}$, namely $\Delta C_{n,y} = \Delta L_{n,y} = 2b$. However, in every kind of transmission line, capacitances only appear in the form $C_{n,y}^{-1}$. Accordingly, the fluctuation of this term is given by $\Delta C_{n,y}^{-1} = \left(\frac{2b}{g}\right)$, where $g = \left(C_{0,y}^2 - b^2\right)$. For $g < 1$ we have $\Delta C_{n,y}^{-1} > \Delta L_{n,y}$, which means that the term $C_{n,y}^{-1}$ introduces a greater disorder into the generic Equation (3) than the disorder introduced by $L_{n,y}$. This fact can induce a decrease in the degree of correlation of the sequence $\{C_n(\alpha)\}$. To compensate this decrease, correlation exponent α must be greater than correlation exponent β of $L_{n,y}(\beta)$ to generate extended states. Consequently, the critical correlation exponents fulfill the condition $\alpha_c \geq \beta_c$, as long as condition $\left(C_{0,y}^2 - b^2\right) < 1$ is valid. Also, for fixed ω, and for given correlation exponents α and β, it is possible to find a critical value of amplitude b of the fluctuation, in such a way that for $b \geq b_c$ all states are localized states (see Figures 10 and 11 of Ref. [88]).

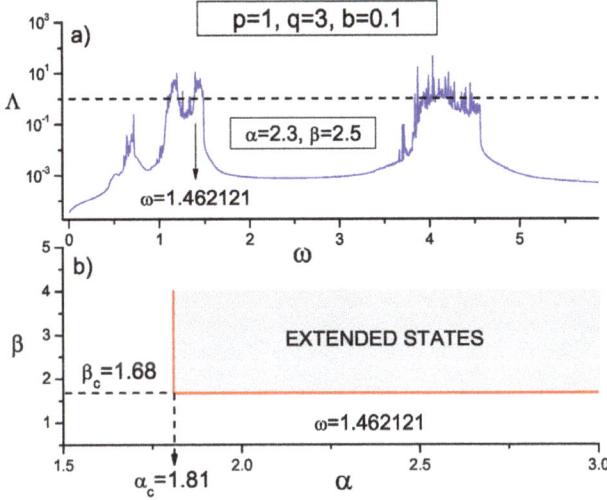

Figure 15. (a) $\Lambda(\omega)$ versus ω, for mixed TL with $p = 1$ and $q = 3$, for $b = 0.1$. We consider two fixed values of the correlation exponents, $\alpha = 2.3$ and $\beta = 2.5$. Only three bands are visible, because $\Lambda(\omega) \to 0$ for the leftmost band (localized states). (b) Phase diagram (α, β) for $\omega = 1.462121$ (indicated by the vertical arrow in (a)). Only for $\alpha \geq \alpha_c = 1.81$ and $\beta \geq \beta_c = 1.68$, with $\alpha_c > \beta_c$ is it possible to find extended states.

The localization behavior of mixed transmission lines with long-range correlated disorder given by (66) can be summarized in Figures 15 and 16, where we studied mixed TL with $p = 1$ and $q = 3$. This case has $d = 4$ bands. Figure 15a shows $\Lambda(\omega)$ versus ω for $b = 0.1$, for two fixed values of the correlation exponents, namely $\alpha = 2.3$ and $\beta = 2.5$. Only three bands are visible, because $\Lambda(\omega) \to 0$ for the leftmost band (localized states). The vertical arrow indicates the specific frequency $\omega = 1.462121$ studied in Figures 15b and 16. For this frequency, Figure 15b shows the phase diagram (α, β). This image indicates that only for $\alpha \geq \alpha_c$ and $\beta \geq \beta_c$, with $\alpha_c > \beta_c$ it is possible to find extended states.

Also, for $\omega = 1.462121$, the critical correlation exponents are $\alpha_c = 1.81$ and $\beta_c = 1.68$. Please note that the asymmetric condition $\alpha_c > \beta_c$ is fulfilled. Figure 16 shows the scaling behavior of the average overlap amplitude $\ln(C_\omega)$ versus $\ln(N)$ for the frequency $\omega = 1.462121$ indicated by vertical arrows in Figure 15a. For fixed $\beta = 3.6$ we obtain the critical value α_c of the correlation exponent, namely $\alpha_c = 1.81$ (see Figure 16a). For $\alpha < \alpha_c$ all states are localized states because we cannot obtain a linear relationship between $\ln(C_\omega)$ and $\ln(N)$. However, for $\alpha \geq \alpha_c$ we only find straight lines with the same slope $m = -1.0$ ($R = 1.0$). This behavior indicates that (NC_ω) is constant for increasing values of N. This is exactly the scaling behavior of the average overlap amplitude for extended states. For fixed $\alpha = 2.0$, Figure 16b shows the same kind of scaling behavior, obtaining $\beta_c = 1.68$.

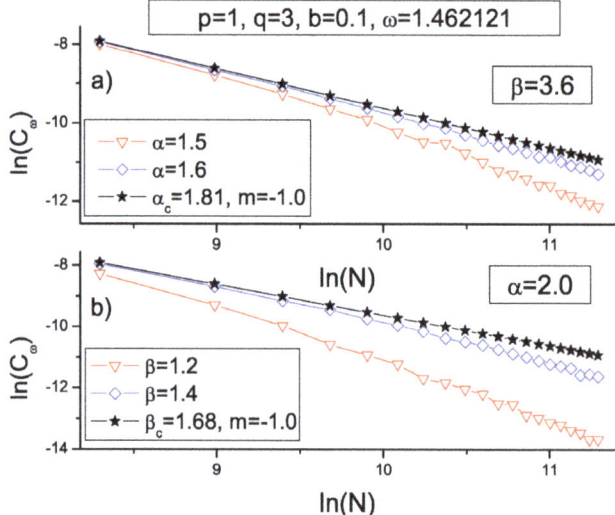

Figure 16. Scaling behavior of $\ln(C_\omega)$ versus $\ln(N)$ for mixed TL with $p = 1$, $q = 3$, $\omega = 1.462121$, and $b = 0.1$. (**a**) For fixed $\beta = 3.6$, only for $\alpha \geq \alpha_c = 1.81$, we find straight lines with the same slope $m = -1.0$ ($R = 1.0$), which indicates an extended behavior. (**b**) For fixed $\alpha = 2.0$, only for $\beta \geq \beta_c = 1.68$, we can obtain linear relationships with $m = -1.0$.

3.3. Diluted Disordered Systems

Hilke [101] introduced the diluted Anderson model, which considers two interpenetrating lattices, i.e., a pure lattice ($\varepsilon_j = \varepsilon_0$), while an Anderson lattice (ε_{jP} is a random number) is periodically distributed with period $P \geq 1$. This means $(P-1)$ diluting elements exist between two Anderson sites. For $P = 1$ we regain the usual Anderson model [1]. This model was generalized [11,12] so that the $(P-1)$ diluting elements are distributed according to a function with certain specific symmetry conditions (see Ref. [11]). The case $\varepsilon_j = \varepsilon_0$ is the most symmetrical of all, and coincides with the results from previous works [101–103]. Depending on the type of symmetry, the dilution process can generate a maximum of up to $(P-1)$ extended states, which are exactly located on some of the edges of the gaps. For constant off-diagonal term, the position of these resonances depends only on the period P and the values of ε_j of the diluting elements. At the same time, resonances are independent of the type of disorder, as well as the degree of correlation in the disordered lattice. In addition, in the resonance, the extended wave function behaves like an intermediate extended function, because its amplitude is zero at each disordered site. The localization behavior of the diluted systems have been studied in the tight-binding quantum case, and in classic systems, like harmonic chains and electric transmission lines [11,12,15,23,80–82,84,96,101–103].

Let us consider the localization behavior of diluted direct transmission lines, with constant capacitances, i.e., $C_n = C_0 \ \forall n$. The inductances L_n, corresponding to disordered sites, have been distributed in different forms. Between two disordered inductances, L_n and L_{n+P}, we put $(P-1)$ identical inductances $L_0 = const.$, where $L_n \neq L_0$. Consequently, the inductances are distributed in the following schematic way ... $L_x \ L_0 \ L_0 \ L_0 \ L_x \ L_0 \ L_0 \ L_0 \ L_x$ ..., where $P = 4$. Because of the full symmetry of the diluting elements, we find exactly $(P-1)$ resonances and $(P-1)$ gaps [11]. For direct transmission lines, resonance frequencies are obtained analytically [80–82]:

$$\begin{aligned} P &= 2, \quad \omega = \sqrt{2}\omega_0 \\ P &= 3, \quad \omega = \sqrt{2 \pm 1}\omega_0 \\ P &= 4, \quad \omega = \left\{\sqrt{2}, \sqrt{2 \pm \sqrt{2}}\right\}\omega_0 \end{aligned} \quad (67)$$

where $\omega_0 = (L_0 C_0)^{-\frac{1}{2}}$ and $L_0 \neq L_n$.

In Ref. [80] the inductances L_n were distributed in (a) a random way, and (b) considering long-range correlated disorder (Fourier Filtering method and Ornstein–Uhlenbeck process). In both cases a continuous distribution of L_n values was used. In addition, in Ref. [81] the inductances L_n were distributed by means of an aperiodic binary sequence of Galois [67], and in Ref. [82] the inductances L_n were distributed considering an asymmetric dichotomous sequence. In all cases studied, the existence of $(P-1)$ intermediate extended states has been demonstrated. Also, the position of the resonance frequencies coincides with theoretical predictions.

On the other hand, the localization behavior of the diluted aperiodic m-tupling distribution of inductances was studied in Ref. [84]. The case $m = 3$ was considered, with $(P-1) = 4$ diluting elements L_0. For numerical calculation, the following data were used: $C_n = C_0 = 0.5 \ \forall n$, $L_A = 1.6$, $L_B = 1.5$ and $L_0 = 1.8$. Figure 6 of Ref. [84] shows (a) the overlap amplitude (NC_ω), and (b) the normalized participation number $\xi(\omega)$. In that picture we can see four gaps and four resonances, which are indicated by vertical dashed lines. Notice that resonances are placed at the left edge of each gap. This result coincides with the theoretical predictions. In Figure 17, we show the average overlap amplitude (NC_ω) for the same case studied in Figure 6 of Ref. [84], but now we study the Thue–Morse sequence, i.e., $m = 2$. Here, we consider four values of P, namely $P = \{1, 2, 3, 4\}$. The case $P = 1$ corresponds to the usual Thue–Morse sequence without dilution. According to (67), for each $P \geq 2$, the frequency of the resonances are: $P = 2$, $\omega = \{1.491\}$; $P = 3$, $\omega = \{1.054, 1.826\}$, and $P = 4$, $\omega = \{0.807, 1.491, 1.948\}$. These theoretical values coincide with the numerical results shown in Figure 17. The same localization behavior of this diluted aperiodic system can be seen in Figure 18 studying the density of states $DOS(\omega)$ versus ω. There we can see that to the left of each gap, the density of states does not fluctuate, which is an indication of the extended nature of the resonance located there. This does not happen on the right side of each gap.

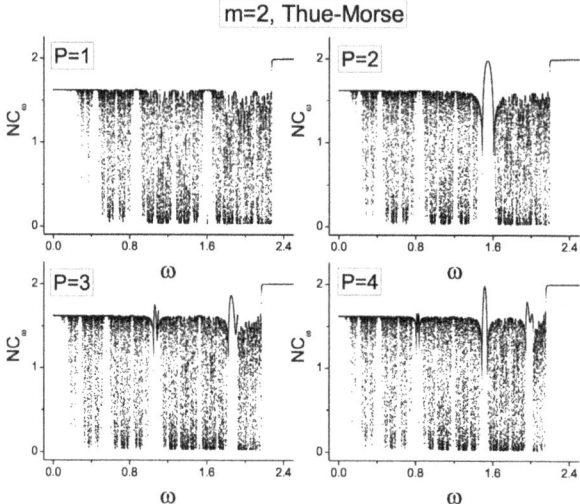

Figure 17. (NC_ω) versus ω, for the Thue–Morse distribution of inductances, with $L_A = 1.6$, $L_B = 1.5$. Four values of the period P are considered, namely $P = \{1, 2, 3, 4\}$. The case $P = 1$ corresponds to the usual Thue–Morse sequence without dilution. For $P \geq 2$, the sequences are diluted with $L_0 = 1.8$, with fixed $C_n = 0.5$. The resonances coincide with the left edge of each of the $(P - 1)$ gaps generated by the dilution process.

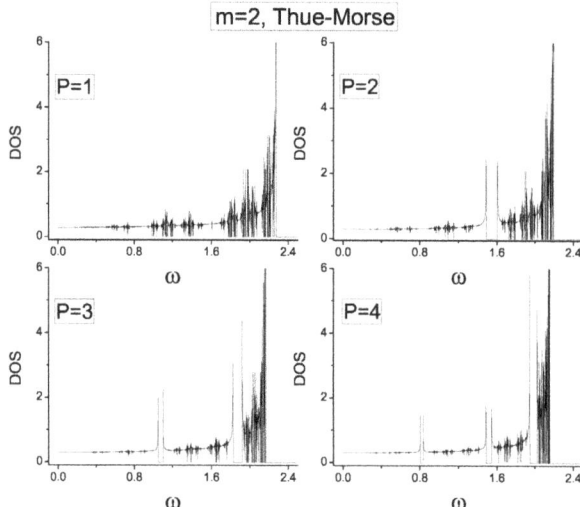

Figure 18. Density of states $DOS(\omega)$ versus ω. To the left of each gap, the density of states does not fluctuate, which is an indication of the extended nature of the resonance located there. The same does not happen on the right side of each gap. In addition, we can see that the localization behavior of the $DOS(\omega)$, is identical to the localization behavior shown by (NC_ω) in Figure 17.

4. Summary and Conclusions

We have presented the results of the study of the localization properties of disordered electrical transmission lines. This study considered three types of TL: dual, direct and mixed. The electrical components of the (capacitances and inductances) were distributed in different non-periodic forms: (a) aperiodic, which included self-similar sequences (Fibonacci and $m-$tupling Thue–Morse), (b) incommensurate sequences (Aubry–André and Soukoulis–Economou), and (c) long-range correlated sequences (binary discrete and continuous). The localization properties of these classical systems were measured using the typical tools used in quantum mechanics to characterize the localization behavior of disordered systems. Specifically, we used the normalized localization length $\Lambda(\omega)$, the inverse participation ratio $IPR(\omega)$, the transmission coefficient $T(\omega)$, the global density of states $DOS(\omega)$, the average overlap amplitude C_ω, and others. Our studies indicate that the localization behavior of classic electric transmission lines is quite similar to the one-dimensional tight-binding quantum model, but at the same time it is possible to observe some significant differences; therefore, it is worth continuing to investigate this type of classical disordered systems.

As a possible application of the study of the localization properties of disordered electric transmission lines, we can consider the neuronal axons that connect two or more neurons through electrical impulses. The axon, which usually is covered by a myelin sheath, can be considered to be a transmission line formed by Schwann cells connected by nodes of Ranvier. It has been established that there exist certain specific genes responsible for stabilizing the internal neuronal structure, which in turn allows the proper transport of the electrical impulse within the axon. The electrical communication between neurons fails if axons are damaged or broken. This can happen in the earliest stages of neurodegenerative diseases or for other reasons. Based on the localization properties of electrical transmission lines studied in this review, it is possible to conclude that electrical communication between neurons prevails, if Schwann cells and Ranvier nodes are distributed in a periodic way or in a very specific aperiodic way. On the contrary, any non-correlated disorder in the axon structure will stop the electrical impulses and the neurons will remain without communication. Consequently, to restore electrical communication between neurons, I can conjecture that the genes responsible for stabilizing the internal neuronal structure have the specific mission of restoring periodicity in the distribution of Schwann cells and Ranvier's nodes.

Up to that point, we have only considered ideal transmission lines, i.e., transmission lines without dissipation (resistance $R = 0$). When we introduced gain ($R_n = -R$, n odd) and loss ($R_n = +R$, n even) balanced pairwise, a $\mathcal{PT}$-symmetric resistive configuration is obtained. For this dissipative system, we can find a critical resistance R_c such that for $R < R_c$ the frequency spectrum is completely real, but for $R > R_c$ the frequency spectrum contains real and complex frequencies. This phenomenon is called a $\mathcal{PT}$-symmetric transition phase, because the TL goes from an unbroken ($R < R_c$) to a broken $\mathcal{PT}$-symmetric phase as a function of resistance R. In addition, we have demonstrated that in the unbroken $\mathcal{PT}$-symmetric phase, the electric current function $I_n(\omega)$ is a symmetric extended function. Conversely, in the broken phase, $I_n(\omega)$ is an antisymmetric localized function. This phase transition was recently found for TL with a very small number of cells considering fixed boundary conditions [89].

In addition to this research, we are currently studying two different lines of research, (a) the influence of non-linear inductances or capacitances in the stability and amplitude of the allowed conducting bands of the unbroken $\mathcal{PT}$-symmetric phase, and (b) the localization behavior of some models of structured transmission lines, in the spirit of the structured systems proposed by Chakrabarti [29]. Specifically, we analyzed TL with a finite number of hanging cells (direct or dual) in random positions in TL. The electric components (capacitances or inductances) of each hanging cell, can contain aperiodic disorder or long-range correlated disorder.

Funding: This research was funded by Dirección de Investigación, Postgrado y Transferencia Tecnológica de la Universidad de Tarapacá, Arica, Chile, grant number 4737-19.

Acknowledgments: The author is grateful for the collaboration of my colleagues and my students who, during the last decade, helped me in the study of the localization properties of disordered electrical transmission lines.

Conflicts of Interest: The author declares no conflict of interest.

Abbreviations

The following abbreviations are used in this manuscript:

TL   Transmission Lines
FFM   Fourier Filtering Method
FFT   Fast Fourier Transform

References

1. Anderson, P.W. Absence of diffusion in certain random lattices. *Phys. Rev.* **1958**, *109*, 1492. [CrossRef]
2. Mott, N.F.; Twose, W.D. The theory of impurity conduction. *Adv. Phys.* **1961**, *10*, 107. [CrossRef]
3. Thouless, D.J. Electrons in disordered systems and the theory of localization. *Phys. Rep.* **1974**, *13*, 93. [CrossRef]
4. Flores, J.C. Transport in models with correlated diagonal and off-diagonal disorder. *J. Phys. Condens. Matter* **1989**, *1*, 8471. [CrossRef]
5. Lifshits, I.M.; Gredeskul, S.A.; Pastur, L.A. *Introduction to the Theory of Disordered Systems*; Wiley: New York, NY, USA, 1989.
6. Dunlap, D.H.; Hu, H.-L.; Philips, P.W. Absence of localization in a random-dimer model. *Phys. Rev. Lett.* **1990**, *65*, 88. [CrossRef]
7. Philips, P.W.; Hu, H.-L. Localization and its absence: A new metallic state for conducting polymers. *Science* **1991**, *252*, 1805. [CrossRef] [PubMed]
8. de Moura, F.A.B.F.; Lyra, M.L. Delocalization in the 1D Anderson model with long-range correlated disorder. *Phys. Rev. Lett.* **1998**, *81*, 3735. [CrossRef]
9. Izrailev, F.M.; Krokhin, A.A. Localization and the mobility edge in one-dimensional potentials with correlated disorder. *Phys. Rev. Lett.* **1999**, *82*, 4062. [CrossRef]
10. Izrailev, F.M.; Krokhin, A.A.; Ulloa, S.E. Mobility edge in aperiodic Kronig-Penney potentials with correlated disorder: Perturbative approach. *Phys. Rev. B* **2001**, *63*, 041102(R). [CrossRef]
11. Lazo, E.; Onell, M.E. Extended states in 1-D Anderson chain diluted by periodic disorder. *Physica B* **2001**, *299*, 173. [CrossRef]
12. Lazo, E.; Onell, M.E. Existence of the delocalized states in two interpenetrated 1-D diluted Anderson chains. *Phys. Lett. A* **2001**, *283*, 376. [CrossRef]
13. Carpena, P.; Galvan, P.B.; Ivanov, P.C.; Stanley, H.E. Metal-insulator transition in chains with correlated disorder. *Nature* **2002**, *418*, 955. [CrossRef] [PubMed]
14. Deych, L.I.; Erementchouk, M.V.; Lisyanky, A.A. Scaling properties of the one-dimensional Anderson model with correlated diagonal disorder. *Phys. Rev. B* **2003**, *67*, 024205. [CrossRef]
15. de Moura, F.A.B.F.; Santos, M.N.B.; Fulco, U.L.; Lyra, M.; Lazo, E.; Onell, M.E. Delocalization and wave-packet dynamics in one-dimensional diluted Anderson models. *Eur. Phys. J. B* **2003**, *36*, 81. [CrossRef]
16. Zhang, W.; Ulloa, S.E. Extended states in disordered systems: role of off-diagonal correlations. *Phys. Rev. B* **2004**, *69*, 153203. [CrossRef]
17. Shima, H.; Nomura, T.; Nakayama, T. Localization-delocalization transition in one-dimensional electron systems with long-range correlated disorder. *Phys. Rev. B* **2004**, *70*, 075116. [CrossRef]
18. Titov, M.; Schomerus, H. Nonuniversality of Anderson localization in short-range correlated disorder. *Phys. Rev. Lett.* **2005**, *95*, 126602. [CrossRef] [PubMed]
19. Izrailev, F.M.; Makarov, N.M. Anomalous transport in low-dimensional systems with correlated disorder. *J. Phys. A* **2005**, *38*, 10613. [CrossRef]
20. Shima, H.; Nakayama, T. Breakdown of Anderson localization in disordered quantum chains. *Microelectr. J.* **2005**, *36*, 422. [CrossRef]

21. de Moura, F.A.B.F.; Malyshev, A.V.; Lyra, M.L.; Malyshev, V.A.; Domínguez-Adame, F. Localization properties of a one-dimensional tight-binding model with nonrandom long-range intersite interactions. *Phys. Rev. B* **2005**, *71*, 174203. [CrossRef]
22. Díaz, E.; Rodriguez, A.; Domínguez-Adame, F.; Malyshev, V.A. Anomalous optical absorption in a random system with scale-free disorder. *Europhys. Lett.* **2005**, *72*, 1018. [CrossRef]
23. Albuquerque, S.S.; de Moura, F.A.B.F.; Lyra, M.L.; Lazo, E. Sensitivity to initial conditions of the wave-packet dynamics in diluted Anderson chains. *Phys. Lett. A* **2006**, *355*, 468. [CrossRef]
24. Esmailpour, A.; Esmaeilzadeh, M.; Faizabadi, E.; Carpena, P.; Reza Rahimi Tabar, M. Metal-insulator transition in random Kronig-Penney superlattices with long-range correlated disorder. *Phys. Rev. B* **2006**, *74*, 024206. [CrossRef]
25. Esmailpour, A.; Cheraghchi, H.; Carpena, P.; Reza Rahimi Tabar, M. Metal–insulator transition in a ternary model with long range correlated disorder. *J. Stat. Mech.* **2007**, P09014. [CrossRef]
26. Kaya, T. Localization-delocalization transition in chains with long-range correlated disorder. *Eur. Phys. J. B* **2007**, *55*, 49. [CrossRef]
27. Benhenni, R.; Senouci, K.; Zekri, N.; Bouamrane, R. Anderson transition in 1D systems with spatial disorder. *Physica A* **2010**, *389*, 1002. [CrossRef]
28. Izrailev, F.M.; Krokhin, A.A.; Makarov, N.M. Anomalous localization in low-dimensional systems with correlated disorder. *Phys. Rep.* **2012**, *512*, 125. [CrossRef]
29. Chakrabarti, A. Electronic states and charge transport in a class of low dimensional structured systems. *Phys. E Low-Dimen. Syst. Nanostruct.* **2019**, *114*, 113616. [CrossRef]
30. Bellani, V.; Diez, E.; Hey, R.; Toni, L.; Tarricone, L.; Parravicini, G.B.; Dominguez-Adame, F.; Gómez-Alcalá, R. Experimental evidence of delocalized states in random dimer superlattices. *Phys. Rev. Lett.* **1999**, *82*, 2159. [CrossRef]
31. Kulh, U.; Izrailev, F.M.; Krokhin, A.A.; Stöckmann, H.-J. Experimental observation of the mobility edge in a waveguide with correlated disorder. *Appl. Phys. Lett.* **2000**, *77*, 633.
32. Krokhin, A.A.; Izrailev, F.M.; Kulh, U.; Stöckmann, H.-J.; Ulloa, S.E. Random 1D structures as filters for electrical and optical signals. *Physica E* **2002**, *13*, 695. [CrossRef]
33. Kulh, U.; Izrailev, F.M.; Krokhin, A.A. Enhancement of localization in lne-dimensional random potentials with long-range correlations. *Phys. Rev. Lett.* **2008**, *100*, 126402.
34. Lugan, P.; Aspect, A.; Sanchez-Palencia, L.; Delande, D.; Grémaud, B.; Muller, C.A.; Miniatura, C. One-dimensional Anderson localization in certain correlated random potentials. *Phys. Rev. A* **2009**, *80*, 023605. [CrossRef]
35. Dietz, O.; Kulh, U.; Stöckmann, H.-J.; Makarov, N.M.; Izrailev, F. Microwave realization of quasi-one-dimensional systems with correlated disorder. *Phys. Rev. B* **2011**, *83*, 134203. [CrossRef]
36. Aubry, S.; André, G. Analyticity breaking and Anderson localization in incommensurate lattices. *Ann. Isr. Phys. Soc.* **1980**, *3*, 133.
37. Kohmoto, M.; Kadanoff, L.P.; Tang, C. Localization Problem in One Dimension: Mapping and Escape. *Phys. Rev. Lett.* **1983**, *50*, 1870. [CrossRef]
38. Ostlund, S.; Pandit, R.; Rand, D.; Schellnhuber, H.J.; Siggia, E.D. One-dimensional Schrödinger equation with an almost periodic potential. *Phys. Rev. Lett.* **1983**, *50*, 1873. [CrossRef]
39. Sokoloff, J.B. Unusual band structure, wave functions and electrical conductance in crystals with incommensurate periodic potentials. *Phys. Rep.* **1985**, *126*, 189. [CrossRef]
40. Kohmoto, M.; Banavar, J.R. Quasiperiodic lattice: Electronic properties, phonon properties, and diffusion. *Phys. Rev. B* **1986**, *34*, 563. [CrossRef]
41. Kohmoto, M. Localization problem and mapping of one-dimensional wave equations in random and quasiperiodic media. *Phys. Rev. B* **1986**, *34*, 5043. [CrossRef]
42. Nori, F.; Rodríguez, J.P. Acoustic and electronic properties of one-dimensional quasicrystals. *Phys. Rev. B* **1986**, *34*, 2207. [CrossRef] [PubMed]
43. Niu, Q.; Nori, F. Renormalization-group study of one-dimensional quasiperiodic systems. *Phys. Rev. Lett.* **1986**, *57*, 2057. [CrossRef] [PubMed]
44. Lu, J.P.; Odagaki, T.; Birman, J.L. Properties of one-dimensional quasilattices. *Phys. Rev. B* **1986**, *33*, 4809. [CrossRef] [PubMed]

45. Fujita, M.; Machita, K. Electrons on one-dimensional quasi-lattices. *Solid State Commun.* **1986**, *59*, 61. [CrossRef]
46. Liu, Y.; Riklund, R. Electronic properties of perfect and nonperfect one-dimensional quasicrystals. *Phys. Rev. B* **1987**, *35*, 6034. [CrossRef] [PubMed]
47. Zaks, M.A.; Pikovsky, A.S.; Kurths, J. On the correlation dimension of the spectral measure for the Thue-Morse sequence. *J. Stat. Phys.* **1987**, *88*, 1387. [CrossRef]
48. Riklund, R.; Severin, M.; Liu, Y. The Thue-Morse aperiodic crystal, a link between the Fibonacci quasicrystal and the periodic crystal. *Int. J. Mod. Phys. B* **1987**, *1*, 121. [CrossRef]
49. Cheng, Z.; Savit, R.; Merlin, R. Structure and electronic properties of Thue-Morse lattices. *Phys. Rev. B* **1988**, *37*, 4375. [CrossRef]
50. Luck, J.M. Cantor spectra and scaling of gap widths in deterministic aperiodic systems. *Phys. Rev. B* **1989**, *39*, 5834. [CrossRef]
51. Kolář, M.; Ali, M.; Nori, F. Generalized Thue-Morse chains and their physical properties. *Phys. Rev. B* **1991**, *43*, 1034. [CrossRef]
52. Ryu, C.S.; Oh, G.Y.; Lee, M.H. Extended and critical wave functions in a Thue-Morse chain. *Phys. Rev. B* **1992**, *46*, 5162. [CrossRef] [PubMed]
53. Huang, D.; Gumbs, G.; Kolar, M. Localization in a one-dimensional Thue-Morse chain. *Phys. Rev. B* **1992**, *46*, 11479. [CrossRef] [PubMed]
54. Ryu, C.S.; Oh, G.Y.; Lee, M.H. Electronic properties of a tight-binding and a Kronig-Penney model of the Thue-Morse chain. *Phys. Rev. B* **1993**, *48*, 132. [CrossRef] [PubMed]
55. Chakrabarti, A.; Karmakar, S.N.; Moitra, R.K. Role of a new type of correlated disorder in extended electronic states in the Thue-Morse lattice. *Phys. Rev. Lett.* **1995**, *74*, 1403. [CrossRef] [PubMed]
56. Deych, L.I.; Zaslavsky, D.; Lisyansky, A.A. Wave localization in generalized Thue-Morse superlattices with disorder. *Phys. Rev. E* **1997**, *56*, 4780. [CrossRef]
57. Oh, G.Y. Existence of Extended Eletronic States in the Thue-Morse Lattice. *J. Korean Phys. Soc.* **1997**, *31*, 808.
58. Oh, G.Y. Quantum dynamics of an electron in a one-dimensional Thue-Morse Lattice. *J. Korean Phys. Soc.* **1998**, *33*, 617.
59. Lazo, E.; Onell, M.E. Localization in one-dimensional systems with generalized Fibonacci disorder. *Revista Mexicana de Física* **1998**, *44*, 52.
60. Maciá, E.; Domínguez-Adame, F. *Electrons, Phonons and Excitons in Low Dimensional Systems*; Editoral Complutense S. A.: Madrid, Spain, 2000.
61. Hui-Fen, G.; Rui-Bao, T. Extended state to localization in random aperiodic chains. *Commun. Theor. Phys.* **2006**, *46*, 929. [CrossRef]
62. Min, N.; Zhixlong, W. A property of m-tuplings morse sequence. *Wuhan Univ. Nat. Sci.* **2006**, *11*, 473. [CrossRef]
63. Maciá, E. The role of periodic order in science and technology. *Rep. Prog. Phys.* **2006**, *22*, 397 [CrossRef]
64. Janssen, T.; Chapuis, G.; de Boissieu, M. *Aperiodic Crystals: From Modulated Phases to Quasicrystals*; Oxford University Press: Oxford, UK, 2007.
65. Steurer, W.; Deloudi, S. *Cristallography of Quasicrystals—Concept, Methods and Structures*; Springer Series in Material Science 126; Springer: Berlin, Germany, 2009.
66. Maciá, E. *Aperiodic Structures in Condensed Matter*; CRC Taylor and Francis: Boca Raton, FL, USA, 2009.
67. Wan, R.; Fu, X. Localization properties of electronic states of one-dimensional Galois sequences. *Solid State Commun.* **2010**, *150*, 919. [CrossRef]
68. Maciá, E. Exploiting aperiodic designs in nanophotonic devices. *Rep. Prog. Phys.* **2012**, *75*, 036502. [CrossRef] [PubMed]
69. Cheng, W.W.; Gong, L.Y.; Shan, C.J.; Sheng, Y.B.; Zhao, S.M. Geometric discord characterize localization transition in the one-dimensional systems. *Eur. Phys. J. D* **2013**, *67*, 121. [CrossRef]
70. Cheng, W.W.; Shan, C.J.; Gong, L.Y.; Zhao, S.M. Measurement-induced disturbance near Anderson localization in one-dimensional systems. *J. Phys. B At. Mol. Opt. Phys.* **2014**, *47*, 175503. [CrossRef]
71. Maciá, E. On the nature of electronic wave function in one-dimensional self-similar and quasiperiodic systems. *ISRN Condens. Matter Phys.* **2014**, *2014*, 165943. [CrossRef]
72. Gong, L.; Li, W.; Zhao, S.; Cheng, W. A measure of localization properties of one-dimensional single electron lattice systems. *Phys. Lett. A* **2016**, *380*, 59. [CrossRef]

73. Maciá, E. Spectral classification of one-dimensional binary aperiodic crystals: An algebraic approach. *Ann. Phys. (Berlin)* **2017**, *2017*, 1700079. [CrossRef]
74. Lambropoulos, K.; Simserides, C. Tight-binding modeling of nucleic acid sequences: Interplay between various types of order or disorder and charge transport. *Symmetry* **2019**, *11*, 968. [CrossRef]
75. Diez, E.; Izrailev, F.; Krokhin, A.; Rodríguez, A. Symmetry-induced tunneling in one-dimensional disordered potentials. *Phys. Rev. B* **2008**, *78*, 035118. [CrossRef]
76. Lazo, E.; Diez, E. Conducting and non-conducting transition in dual transmission lines using a ternary model with long-range correlated disorder. *Phys. Lett. A* **2010**, *374*, 3538. [CrossRef]
77. Lazo, E.; Diez, E. Conducting properties of classical transmission lines with Ornstein-Uhlenbeck type disorder. *Phy. Lett. A* **2011**, *375*, 2122. [CrossRef]
78. Lazo, E.; Mellado, F.; Saavedra, E. Rényi entropies of electrical transmission lines with Fibonacci distribution of inductances. *Phys Lett. A* **2012**, *376*, 3423. [CrossRef]
79. Lazo, E.; Diez, E. Phase transition in transmission lines with long-range correlated disorder. *Physica B* **2013**, *419*, 19. [CrossRef]
80. Lazo, E. Generation of intermediate states in diluted disordered direct transmission lines. *Physica B* **2014**, *432*, 121. [CrossRef]
81. Lazo, E.; Humire, F.R.; Saavedra, E. Generation of extended states in diluted transmission lines with distribution of inductances according to Galois sequences: Hamiltonian map approach. *Physica B* **2014**, *452*, 74. [CrossRef]
82. Lazo, E.; Humire, F.R.; Saavedra, E. Disorder-order transitions in diluted and nondiluted direct transmission lines with asymmetric dichotomous noise. *Int. J. Mod. Phys. C* **2014**, *25*, 1450023. [CrossRef]
83. Lazo, E.; Saavedra, E.; Humire, F.R.; Castro, C.E.; Cortés-Cortxexs, F. Localization properties of transmission lines with generalized Thue-Morse distribution of inductances. *Eur. Phys. J. B* **2015**, *88*, 216. [CrossRef]
84. Lazo, E.; Castro, C.E.; Cortés-Cortxexs, F. Overlap amplitude and localization properties in aperiodic diluted and non-diluted direct electric transmission lines. *Phys. Lett. A* **2016**, *380*, 3284. [CrossRef]
85. Lazo, E.; Garrido, A.; Neira, F. The effect of non-linear capacitances in the localization properties of aperiodic dual electric transmission lines. *Eur. Phys. J. B* **2016**, *89*, 249. [CrossRef]
86. Lazo, E. Accurate measurement of the localization properties of electric transmission lines using the overlap amplitude. *Eur. Phys. J. D* **2017**, *71*, 144. [CrossRef]
87. Lazo, E.; Cortés-Cortxexs, F. Disordered mixed transmission lines: Localization behavior. *Eur. Phys. J. Plus* **2019**, *134*, 28. [CrossRef]
88. Lazo, E. Influence of two independent sources with long-range correlated disorder in the localization properties of mixed transmission lines. *Phys.-Low-Dimens. Syst. Nanostruct.* **2019**, *114*, 113628. [CrossRef]
89. Humire, F.R.; Lazo, E. $\mathcal{PT}$-symmetric direct electrical transmission lines: Localization behavior. *Phys. Rev.* **2019**, *100*, 022221. [CrossRef] [PubMed]
90. Lazo, E. Spectral properties of a chain of polyhedra. *Phys. Status Solidi B* **1992**, *169*, 359. [CrossRef]
91. Lazo, E. Mixed Crystal electronic structure of $MS_{3-x}Se_x$. *Phys. Status Solidi B* **1992**, *170*, 463. [CrossRef]
92. Rényi, A. On the foundations of information theory. *Rev. Int. Stat.* **1965**, *33*, 1. [CrossRef]
93. Coffman, V.; Kundu, J.; Wooters, W.K. Distributed entanglement. *Phys. Rev. A* **2000**, *61*, 052306. [CrossRef]
94. Lakshminarayan, A.; Subrahmanyam, V. Entanglement sharing in one-particle states. *Phys. Rev. A* **2003**, *67*, 052304. [CrossRef]
95. Li, H.; Wang, X.; Hu, B. Bipartite entanglement and localization of one-particle states. *J. Phys. A Math. Gen.* **2004**, *37*, 10665. [CrossRef]
96. Albuquerque, S.S.; de Moura, F.A.B.F.; Lyra, M.L. Vibrational modes in harmonic chains with diluted disorder. *Physica A* **2005**, *357*, 165. [CrossRef]
97. de Moura, F.A.B.F.; Viana, L.P.; Frery, A.C. Vibrational modes in aperiodic one-dimensional harmonic chains. *Phys. Rev. B* **2006**, *73*, 212302. [CrossRef]
98. Albuquerque, S.S.; Santos, J.L.L.D.; de Moura, F.A.B.F.; Lyra, M.L. Enhanced localization, energy anomalous diffusion and resonant mode in harmonic chains with correlated mass-spring disorder. *J. Phys. Condens. Matter* **2015**, *27*, 175401. [CrossRef] [PubMed]
99. Sales, M.O.; Alburquerque, S.S.; de Moura, F.A.B.F. Energy transport in a one-dimensional harmonic ternary chain with Ornstein–Uhlenbeck disorder. *J. Phys. Condens. Matter* **2012**, *24*, 495401. [CrossRef] [PubMed]

100. Kaya, T. One-dimensional Anderson model with dichotomic correlation. *Eur. Phys. J. B* **2007**, *60*, 313. [CrossRef]
101. Hilke, M. Localization properties of the periodic random Anderson model. *J. Phys. A Mater. Gen.* **1997**, *30*, L367. [CrossRef]
102. Domínguez-Adame, F.; Gómez, I.; Avakyan, A.; Sedrakyan, D.; Sedrakyan, A. Electron States in a Class of One-Dimensional Random Binary Alloys. *Phys. Status Solidi B* **2000**, *221*, 633. [CrossRef]
103. Deng, W. Anomalous Anderson localization. *Physica B* **2000**, *279*, 224. [CrossRef]

© 2019 by the author. Licensee MDPI, Basel, Switzerland. This article is an open access article distributed under the terms and conditions of the Creative Commons Attribution (CC BY) license (http://creativecommons.org/licenses/by/4.0/).

Review

Tight-Binding Modeling of Nucleic Acid Sequences: Interplay between Various Types of Order or Disorder and Charge Transport

Konstantinos Lambropoulos * and Constantinos Simserides *

Department of Physics, National and Kapodistrian University of Athens, Panepistimiopolis, Zografos, 15784 Athens, Greece
* Correspondence: klambro@phys.uoa.gr (K.L.); csimseri@phys.uoa.gr (C.S.)

Received: 30 June 2019; Accepted: 11 July 2019; Published: 1 August 2019

Abstract: This review is devoted to tight-binding (TB) modeling of nucleic acid sequences like DNA and RNA. It addresses how various types of order (periodic, quasiperiodic, fractal) or disorder (diagonal, non-diagonal, random, methylation et cetera) affect charge transport. We include an introduction to TB and a discussion of its various submodels [wire, ladder, extended ladder, fishbone (wire), fishbone ladder] and of the process of renormalization. We proceed to a discussion of aperiodicity, quasicrystals and the mathematics of aperiodic substitutional sequences: primitive substitutions, Perron–Frobenius eigenvalue, induced substitutions, and Pisot property. We discuss the energy structure of nucleic acid wires, the coupling to the leads, the transmission coefficients and the current–voltage curves. We also summarize efforts aiming to examine the potentiality to utilize the charge transport characteristics of nucleic acids as a tool to probe several diseases or disorders.

Keywords: nucleic acids; aperiodic; quasiperiodic; fractal; order; disorder; energy structure; charge transport

1. Introduction

Nucleic acids are polymeric macromolecules consisting of units that are called nucleotides. The term nucleic acids is the overall name of deoxyribonucleic acid (DNA) and ribonucleic acid (RNA). DNA's nucleotide sequence carries the genetic instructions for the development, function, growth and reproduction of living organisms and several viruses. Although RNA's primary role is to carry out the instructions encoded in DNA for protein synthesis, it also acts like a catalyst of biochemical reactions, while it is the genetic material of many viruses.

For more than sixty years now, the double-stranded structure of DNA has been known [1]. The nucleotides of each strand are composed of one of four planar, aromatic, nitrogenous bases, i.e., guanine (G), cytosine (C), adenine (A) or thymine (T), a pentose sugar (deoxyribose), and a phosphate group. Covalent, phosphodiester bonds between pentoses and phosphate groups of adjacent nucleotides form an alternating sugar-phosphate backbone. The purines (G or A) of a nucleotide belonging to a strand are joined together with the pyrimidines of the other strand (C or T, respectively) via (three or two, respectively) hydrogen bonds, forming the double helix structure. This specificity in the way bases match ensures that G is always bonded with C, and A is always bonded with T. Pairing between non-complementary bases results in mutations that can be detrimental to the development of an organism. In RNA, deoxyribose (whose 2-carbon is bonded with a hydrogen) is replaced by ribose (whose 2-carbon is bonded with a hydroxyl group), and T is replaced by uracil (U). Furthermore, RNA molecules are single-stranded; however, some viruses possess double-stranded RNA (other viruses can contain even single-stranded DNA).

Although the study of nucleic acids is mainly associated with molecular biology and genetics, today, a broad interdisciplinary community is interested in biological systems, such as nucleic acids and analogues. The base-pair stack of nucleic acids creates a nearly one-dimensional π-stack that allows charge carrier movement, i.e., charge transfer and transport. Let us distinguish between these two terms: transfer means that a carrier, created or injected at a specific nucleotide, moves to a more favorable location, while transport implies the use of electrodes and the application of external voltage between them. Charge transfer is the basis of many biological processes, e.g., in various proteins [2] including metalloproteins [3], and enzymes [4], with medical and bioengineering applications [5,6], while it plays a role in DNA damage and repair [7–9]. Charge transport might be an indicator to distinguish pathogenic from non-pathogenic mutations at an early stage [10].

From a physicist's point of view, the charge transfer and transport properties of nucleic acids are studied in order to obtain a deeper understanding of their biological functions as well as for potential applications, such as nanosensors, nanocircuits or molecular wires, due to their high yield synthesis, near-unity purification, and nanoscale self-organization [11–13]. There are many external (aqueousness, presence of counterions, extraction process, electrodes, contacts, purity, substrate), and internal (such as the base-pair sequence and geometry) factors that affect carrier motion along nucleic acids. Both ab initio calculations [14–22] and model Hamiltonians [23–34] have been used to theoretically explore the variety of experimental results that predict electrical behavior ranging from metallic to insulating, as well as the underlying mechanisms.

It has become evident that the influence of various types of order or disorder plays a central role in the energy structure and the charge transport properties of nucleic acids. This interplay between various types of order or disorder and charge transport is addressed in this brief review. This is done in the context of one of the most widely applied theoretical methods, i.e., with Tight-Binding (TB), because of its simplicity and low computational cost.

The rest of this review is organized as follows. In Section 2, we present the TB formulation and explain some of its most common variations applied in the literature for the study of nucleic acids. In Section 3, we overview several aperiodic substitutional sequences that highlight the influence of disorder in the properties of nucleic acids. In Section 4, we discuss the energy spectra of ordered and disordered nucleic acid sequences. In Section 5, we focus on electron transmission and on the influence of coupling the examined systems with leads. Section 6 is dedicated to the influence of various types of order or disorder on the current–voltage ($I - V$) curves of nucleic acids. Finally, in Section 7, we make some concluding remarks.

2. Tight-Binding and Its Application in Nucleic Acids

TB is an approximate method widely used in condensed matter physics to determine the electronic structure of a solid through the expansion of its wavefunction as a superposition of the wavefunctions corresponding to the isolated moieties located at each lattice site [35]. As the name of the model suggests, the main hypothesis in TB is that the system's orbitals are tightly bound at the sites at which they belong, so that the overlap with neighboring orbitals is small. Hence, the electronic wavefunction of the moiety that occupies a lattice site is rather similar to the orbital of the free moiety. As a result, the corresponding energy of the electron will be rather close to the (negative) ionization energy of the free moiety due to the weak interaction with its neighbors. This picture is applicable at the bands formed by the core electrons of metals, the valence and conduction bands of insulators and semiconductors, as well as the valence and conduction bands arising from localized d or f states (e.g., in transition metals and rare earths).

Today, several decades after its introduction [36], TB has evolved into a fast and efficient approach, employable to numerous problems regarding the electronic structure and properties of matter, requiring various degrees of accuracy [37,38]. Its main advantages include its intuitive simplicity, the ability it gives to obtain analytic results in several cases, and its low computational cost [39]. The latter makes TB applicable to large systems, currently unreachable by the more sophisticated ab initio methods, such as Density Functional Theory (DFT). In contrast to those methods, TB is semi-empirical, in the sense that an external set of parameters is needed in order to perform calculations. These parameters are (a) the on-site energies that correspond to the energy of the electrons that belong to each lattice site, and (b) the hopping (or transfer) integrals that correspond to the coupling of orbitals which belong to neighboring sites.

Over the last few decades, TB has been widely used to describe, among others, polymers and organic systems. One-dimensional TB models are commonly applied to study the energy structure and thermal, magnetic as well as charge transfer and transport properties of π-conjugated organic systems that are candidates for molecular wires, such as nucleic acids and analogues. Those models have varying degrees of complexity, and each one of them requires a different number of parameters. As far as nucleic acids are concerned, the models employed include, inter alia, the Wire Model (WM), the Ladder Model (LM), the Extended Ladder Model (ELM), the Fishbone Model (FM) and the Fishbone Ladder Model (FLM). Generally, the studied systems consist of N monomers extended at L chains ($L \ll N$, since nucleic acids are approximately one-dimensional). The problem is reduced to the solution of the so-called system of TB equations, which is a system of coupled stationary, algebraic equations or differential equations of first order, equivalent to a discretized form of the time-independent or time-dependent Schrödinger equation. As far as nucleic acids are concerned, the stationary TB system of equations can be compactly written in the matrix form

$$E\vec{\Psi}_n = \varepsilon_n \vec{\Psi}_n + \tau_{n-1}^T \vec{\Psi}_{n-1} + \tau_n \vec{\Psi}_{n+1}, \tag{1}$$

for $n = 1, 2, \ldots, N$. $\vec{\Psi}_n$ is a vector matrix containing the elements of the wavefunction that correspond to monomer n, i.e., $\vec{\Psi}_n = (\psi_n^1\ \psi_n^2\ \ldots\ \psi_n^L)^T$, ε_n is a symmetric $L \times L$ matrix containing the on-site energies of each site, ϵ_n^l and the hopping integrals $t_n^{ll'}$ between the sites of the monomer that belong to different chains, and τ_n is a generally non-symmetric $L \times L$ matrix containing the hopping integrals $t_{nn'}^{ll'}$ between each site of a monomer and the neighboring sites of the next monomer. Finally, E is the energy. The situation is schematically presented in Figure 1. From Bloch's theorem, it holds that $\vec{\Psi}_{N+n} = z\vec{\Psi}_n$, where z generally lies in the unit circle ($z = z^* = 1$, for cyclic boundaries, or $z = z^* = 0$ for fixed boundaries). Hence, the solution of the system of Equation (1) can be reduced to the diagonalization of the Hamiltonian matrix, written in block form as

$$\mathbf{H} = \begin{pmatrix} \varepsilon_1 & \tau_1 & & & z^*\tau_0^T \\ \tau_1^T & \varepsilon_2 & \tau_2 & & \\ & \tau_2^T & \varepsilon_3 & \tau_3 & \\ & & \ddots & \ddots & \ddots \\ z\tau_N & & & \tau_{N-1}^T & \varepsilon_N \end{pmatrix}. \tag{2}$$

Equivalently, Equation (1) can be written in the form

$$\begin{pmatrix} \vec{\Psi}_{n+1} \\ \vec{\Psi}_n \end{pmatrix} = \begin{pmatrix} \tau_n^{-1}(E - \varepsilon_n) & -\tau_n^{-1}\tau_{n-1}^T \\ \mathbf{1} & \mathbf{0} \end{pmatrix} \begin{pmatrix} \vec{\Psi}_n \\ \vec{\Psi}_{n-1} \end{pmatrix} = \mathbf{Q}_n(E) \begin{pmatrix} \vec{\Psi}_n \\ \vec{\Psi}_{n-1} \end{pmatrix}, \tag{3}$$

where $\mathbf{Q}_n(E)$ is called the transfer matrix of monomer n, and $\mathbf{1}, \mathbf{0}$ are the unit and zero matrix of order L. The product

$$\mathbf{M}_N(E) = \prod_{n=N}^{1} \mathbf{Q}_n(E) \tag{4}$$

defines the global transfer matrix of the system, which satisfies the relation,

$$\mathbf{M}_N(E) \begin{pmatrix} \vec{\Psi}_1 \\ \vec{\Psi}_0 \end{pmatrix} = \begin{pmatrix} \vec{\Psi}_{N+1} \\ \vec{\Psi}_N \end{pmatrix} = z \begin{pmatrix} \vec{\Psi}_1 \\ \vec{\Psi}_0 \end{pmatrix}, \tag{5}$$

and contains all the information about its energetics. In fact, since z is an eigenvalue of the global transfer matrix, with eigenvector $\left(\vec{\Psi}_1 \; \vec{\Psi}_0\right)^T$, the whole eigenvector of the Hamiltonian matrix of Equation (2) can be reconstructed via a successive application of Equation (3) [40,41]. Hence, when z is an eigenvalue of $\mathbf{M}_N(E)$, E is an eigenvalue of the system's Hamiltonian. Thus, both methods can be used to determine the energy structure of the system. The form of the matrices in Equation (1) for various TB models is presented in Table 1. Some details on each of these TB models are discussed below.

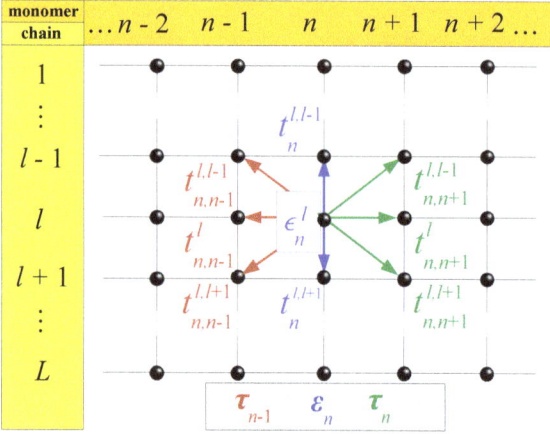

Figure 1. Schematic representation of a TB model consisting of N monomers, extended at L chains. Within the model, we take into account (a) the on-site energies of each site, ϵ_n^l, and the inter-chain hopping integrals, $t_n^{l,l'}$, i.e., between the sites of the monomer (blue), as well as (b) the inter-monomer hopping integrals, $t_{n,n'}^{l,l'}$, i.e., between each site of a monomer and the neighboring sites of the previous (red) and the next (green) monomers. The former are contained in the matrix ε_n, while the latter in the matrices τ_{n-1} and τ_n, respectively.

Table 1. Form of the matrices $\vec{\Psi}_n$, ε_n, τ_n in the TB system of equations (Equation (1)) for several models used to describe nucleic acids and analogues: the Wire Model (WM), the Ladder Model (LM), the Extended Ladder Model (ELM), the Fishbone Model (FM) and the Fishbone Ladder Model (FLM).

Model	L	$\vec{\Psi}_n$	ε_n	τ_n
WM	1	ψ_n	ϵ_n	$t_{n,n+1}$
LM	2	$\begin{pmatrix} \psi_n^1 \\ \psi_n^2 \end{pmatrix}$	$\begin{pmatrix} \epsilon_n^1 & t_n^{1,2} \\ t_n^{2,1} & \epsilon_n^2 \end{pmatrix}$	$\begin{pmatrix} t_{n,n+1}^{1,1} & 0 \\ 0 & t_{n,n+1}^{2,2} \end{pmatrix}$
ELM	2	$\begin{pmatrix} \psi_n^1 \\ \psi_n^2 \end{pmatrix}$	$\begin{pmatrix} \epsilon_n^1 & t_n^{1,2} \\ t_n^{2,1} & \epsilon_n^2 \end{pmatrix}$	$\begin{pmatrix} t_{n,n+1}^{1,1} & t_{n,n+1}^{1,2} \\ t_{n,n+1}^{2,1} & t_{n,n+1}^{2,2} \end{pmatrix}$
FM	3	$\begin{pmatrix} \psi_n^1 \\ \psi_n^2 \\ \psi_n^3 \end{pmatrix}$	$\begin{pmatrix} \epsilon_n^1 & t_n^{1,2} & 0 \\ t_n^{2,1} & \epsilon_n^2 & t_n^{2,3} \\ 0 & t_n^{3,2} & \epsilon_n^3 \end{pmatrix}$	$\begin{pmatrix} 0 & 0 & 0 \\ 0 & t_{n,n+1}^{2,2} & 0 \\ 0 & 0 & 0 \end{pmatrix}$
FLM	4	$\begin{pmatrix} \psi_n^1 \\ \psi_n^2 \\ \psi_n^3 \\ \psi_n^4 \end{pmatrix}$	$\begin{pmatrix} \epsilon_n^1 & t_n^{1,2} & 0 & 0 \\ t_n^{2,1} & \epsilon_n^2 & t_n^{2,3} & 0 \\ 0 & t_n^{3,2} & \epsilon_n^3 & t_n^{3,4} \\ 0 & 0 & t_n^{4,3} & \epsilon_n^4 \end{pmatrix}$	$\begin{pmatrix} 0 & 0 & 0 & 0 \\ 0 & t_{n,n+1}^{2,2} & 0 & 0 \\ 0 & 0 & t_{n,n+1}^{3,3} & 0 \\ 0 & 0 & 0 & 0 \end{pmatrix}$

2.1. Wire Model

WM is the simplest TB model to describe nucleic acids and analogues [42,43]. It can be applied to mimic either single-stranded nucleic acids and hairpins at the single-base level [44] or double-stranded ones [45] at the base-pair level. In other words, if the WM refers to a single-stranded nucleic acid, then the on-site energies are related to the energy levels of the four possible bases and the hopping integrals to the interaction between bases, while, if it refers to a double-stranded nucleic acid, then the on-site energies are related to the energy levels of the two possible base-pairs (incorporating the hydrogen bonding) and the hopping integrals to the interaction between base-pairs. It consists of just one chain ($L = 1$) and the parameters needed for its employment are the on-site energies of the bases or base pairs, ϵ_n, and the hopping integrals between successive bases or base pairs, t_n. A schematic representation of the WM is shown in Figure 2a.

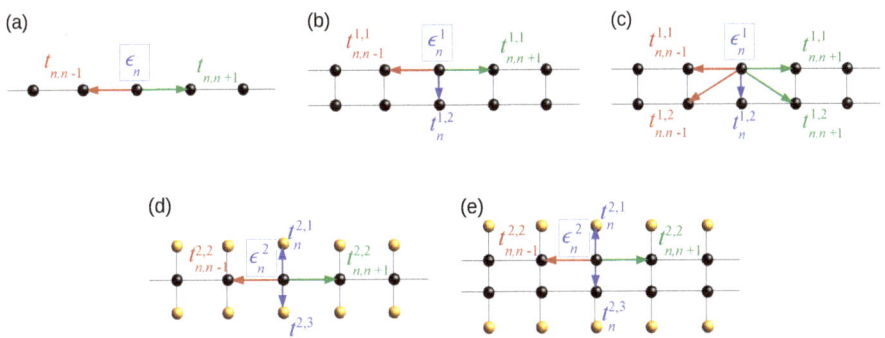

Figure 2. Schematic representation of the TB models listed in Table 1. (**a**) Wire Model (WM); (**b**) Ladder Model (LM); (**c**) Extended Ladder Model (ELM); (**d**) Fishbone Model (FM); (**e**) Fishbone Ladder Model (FLM).

2.2. Ladder Model

LM is the simplest model that can address the influence of base-pairing in the energetics of nucleic acids [42,46]. It consists of two chains ($L = 2$) and the parameters needed for its employment are the on-site energies of the bases, ϵ_n^l, the inter-strand hopping integrals between successive bases, $t_{n,n\pm1}^{ll}$, and the intra-base-pair hopping integrals, $t_n^{ll'}$, due to the hydrogen bonds formed by the complementary bases in a pair. A schematic representation of the LM is shown in Figure 2b.

2.3. Extended Ladder Model

ELM is a more detailed version of the LM, including the inter-strand hopping integrals, $t_{n,n\pm1}^{ll'}$, between the bases of successive base pairs [46,47]. A schematic representation of the ELM is shown in Figure 2c.

2.4. Fishbone Model

FM is the simplest model that can take into account the effect of the sugar-phosphate backbone [29,42]. It consists of three chains ($L = 3$). The central one corresponds to the base pairs, with each one being interconnected with the top and bottom chains, which represent the backbone sites. The latter are not connected with each other, since the insulating sugars are separating phosphate groups from one another [11,48]. Hence, the parameters needed for its employment are the on-site energies, ϵ_n^l, of the base pairs ($l = 2$) and the backbone sites ($l = 1, 3$), the intra-strand hopping integrals between successive base pairs, $t_{n,n\pm1}^{2,2}$, and the inter-strand hopping integrals, $t_n^{ll'}$, between the base pairs and the backbone. A schematic representation of the FM is shown in Figure 2d.

2.5. Fishbone Ladder Model

FLM is a combination of the LM and the FM [29,42]. It thus includes both the effect of base-pairing and the presence of the sugar-phosphate backbone. It consists of four chains ($L = 4$). The two central ones ($l = 2, 3$) correspond to the nitrogenous bases and the edge ones ($l = 1, 4$) to the backbone sites. Hence, the parameters needed for its employment are the on-site energies, ϵ_n^l, of the base pairs ($l = 2, 3$) and the backbone ($l = 1, 4$), the intra-strand hopping integrals between base pairs, $t_{n,n\pm 1}^{ll}$ ($l = 2, 3$) and the inter-strand hopping integrals between the bases of a base pair as well as between each base and the backbone, $t_n^{ll'}$. A schematic representation of the FLM is shown in Figure 2e.

2.6. Additional Remarks

Apart from the models described above, one can introduce several other variants to describe nucleic acids. For example, an obvious extension would be a fishbone extended ladder model. Additionally, several other models have been proposed, including intra-backbone interactions [27,46,49], single-stranded nucleic acids with a backbone [49] and explicit inclusion of helicity [50] strain [51], and spin–orbit coupling [52] effects. We also mention that more complex models can be reduced to simpler ones via a renormalization scheme, which reduces the degrees of freedom of the system. Then, the on-site energies of the renormalized Hamiltonian are energy-dependent. This procedure is important when environmentally induced effects are considered [29]. For example, the FLM can be reduced into an LM via a one-step renormalization procedure [53], or to an even simpler WM via a two-step renormalization procedure [54,55].

Several techniques can be applied to solve the models, depending on what is studied, such as the numerical diagonalization of the Hamiltonian in Equation (2) [47,56,57], the transfer matrix method [58–60] outlined above, and the Non-Equilibrium Green's Function technique [61]. As it is apparent from Equation (3), the transfer matrix method is not applicable if the matrices τ_n are singular. Generally, this is the case, e.g., for the FM and the FLM (cf. Table 1). Then, a renormalization scheme is needed to apply the transfer matrix method.

Relevant parametrizations for nucleic acids have been proposed in many works and used within various TB models. For example, for on-site energies and hopping integrals, cf. Refs. [16,17,20,62,63], for on-site energies, cf. Refs. [64–70], and for hopping integrals, cf. Refs. [71–73]. Such parametrizations allow researchers to go beyond the chemically unrealistic treatments, such as the assumptions that all hopping integrals or on-site energies are equal, i.e., disorder in the Hamiltonian is either purely diagonal or off-diagonal, respectively, and address in more detail the complexity of nucleic acid energy structure.

3. Aperiodic One-Dimensional Wires

The dichotomy between the notions of order and disorder has expanded beyond a simple distinction between periodicity and aperiodicity, since the first observation of icosahedral diffraction patterns in the spectrum of an $Al_{0.86}Mn_{0.14}$ alloys [74] (2011 Nobel Prize in Chemistry for Prof. Dan Shechtman). The discussion that opened in the scientific community following this and other relevant discoveries led to a change in the very definition of the term crystal by the International Union of Crystallography in 1992, expanding it from referring solely to periodically arranged structures to "any solid having an essentially discrete diffraction diagram" [75]. This extended notion of crystals encompasses a whole family of structures, called quasi-periodic crystals or quasicrystals. Quasicrystals do not possess the translation symmetry that is inherent to classical (periodic) crystals; however, they possess inflation/deflation symmetry which leads to long-range order as well.

The discovery of quasicrystals has turned scientific interest into the study of specific one-dimensional aperiodic lattices, modeled with TB [76], i.e., described by Equation (1). The lattices are typically created using substitutional sequences. Apart from the interest the study of such systems has in itself, it is applicable, among other systems of physical relevance, in nucleic acids, as seen in Section 2. The ability to produce synthetic, de novo, nucleic acid sequences of interest [77], using mainly the phosphoramidite method [78] (although other promising methods have recently been proposed [79]), provides a chance not only to examine theoretical predictions regarding aperiodic structures, but also to create molecular wires with tailored properties. Below, we present some details about substitutional sequences as well as some of the most commonly used ones in the literature of one-dimensional wires generally, and nucleic acids specifically.

3.1. Aperiodic Substitutional Sequences

Aperiodic substitutional sequences are based on an alphabet, e.g., $\mathcal{A} = \{A, B, C, D, \dots\}$ equipped with substitution rules that apply to each of its letters, $s(j), \forall j \in \mathcal{A}$. In the case of nucleic acids, the alphabet letters correspond to nitrogenous bases, i.e., G, C, A, T, U (for double-stranded chains the complementary strand is implied). The sequences start with a seed, i.e., a letter belonging to the alphabet (0th generation of the sequence). The substitution rules replace each alphabet letter by finite words consisting of alphabet letters, i.e., $s(j) = j'_1 j'_2, \dots j'_k, \forall j \in \mathcal{A}$. Iterating this procedure g times constructs the gth generation of the sequence.

Substitutional sequences can, in most cases, be described by introducing the substitution matrix, $\mathbf{S}$. It is a square, non-negative matrix of order $card(\mathcal{A})$ (the cardinality of a set is the number of elements of the set), and its elements are $S_{ij} = n_i[s(j)]$, where $n_i[s(j)]$ is the number of times the letter i is present in the substitution rule $s(j)$. Notice that, by definition, $\mathbf{S}$ does not contain information about the ordering of letters in the sequence, hence more than one substitutions can have the same substitution matrix. However, the substitution matrix reveals much information about the underlying order and other properties of the corresponding sequence at the thermodynamic limit.

3.2. Primitive Substitutions and the Perron–Frobenius Eigenvalue

The matrix $\mathbf{S}$ (and, hence, the substitution) is called primitive if there exists a natural number k such that S^k is a positive matrix. For primitive substitutions, the Perron–Frobenius theorem [80,81] guarantees that $\mathbf{S}$ has a largest, unique, real, positive eigenvalue, λ_{PF}, and its corresponding (left and right) eigenvectors can be chosen to have strictly positive entries. The components of the right eigenvector associated with λ_{PF}, normalized such as their sum is unity, give the asymptotic relative frequencies of the letters in $\mathcal{A}$. Hence, using $\mathbf{S}$, one can determine the occurrence percentage of each nucleotide in a substitutional nucleic acid sequence.

3.3. Induced Substitutions

In addition to the previous discussion, it is also possible to determine the letter frequencies of the legal words of length k in a substitutional sequence with primitive $\mathbf{S}$ (corresponding to nucleotide k-plets). This can be done as follows [82]; let $W = \{w = j_1 j_2 \dots j_k, \forall j \in \mathcal{A}\}$ be the set of the legal k-letter words in the sequence and $s(w) = s(j_1)s(j_2)\dots s(j_k) = j'_1 j'_2 \dots j'_n$ the word constructed from a letter-by-letter substitution of the word w. Then, the induced substitution of a k-letter word, $s_k(w) = (j'_1 j'_2 \dots j'_k)(j'_2 j'_3 \dots j'_{k+1}) \dots (j'_l j'_{l+1} \dots j'_{l+k-1})$, where l is the number of letters in $s(j_1)$, is also primitive. Hence, an induced primitive substitution matrix $\mathbf{S}_k$ can be defined, from which the asymptotic letter frequencies of the legal k-letter words of the sequence can be determined using the Perron–Frobenius theorem. For sequences in which $\mathbf{S}$ is defined via a helping alphabet [83], k-letter word frequencies can be deduced in the same fashion from the legal $2k$-letter words of the helping alphabet.

3.4. The Pisot Property

A real algebraic integer (i.e., a real solution of a monic integer polynomial) is said to be a Pisot–Vijayaraghavan number if its modulus is larger than unity, and all its algebraic conjugates (i.e., the other solutions of the polynomial) have modulus strictly less than unity [84]. A substitution has the Pisot property if the matrix $\mathbf{S}$ has a largest, unique, real, positive eigenvalue which is a Pisot–Vijayaraghavan number, and for all the other eigenvalues, λ, it holds that $|\lambda| < 1$. If the characteristic polynomial of $\mathbf{S}$ is irreducible over the rationals, the Pisot substitution is called irreducible. Irreducible Pisot substitutions are a subclass of primitive substitutions [85].

Let us remember some definitions. Given n linearly independent vectors $\boldsymbol{b}_1, \boldsymbol{b}_2, \dots \boldsymbol{b}_n \in \mathbb{R}^m$, the lattice generated by them is defined as $\mathcal{L}(\boldsymbol{b}_1, \boldsymbol{b}_2, \dots \boldsymbol{b}_n) = \sum_i x_i \boldsymbol{b}_i, x_i \in \mathbb{Z}$. We call the set $\boldsymbol{b}_1, \boldsymbol{b}_2, \dots \boldsymbol{b}_n$

a *basis* of the lattice. We say that the *rank* of the lattice is n and its *dimension* is m. The Fourier transform of the (direct) lattice is a lattice that is called the reciprocal lattice.

Furthermore, according to the Lebesgue's decomposition theorem, any measure on $\mathbb{R}$ can be decomposed into three parts: a pure point (or discrete) part, an absolutely continuous part, and singularly continuous part. This theorem helps to categorize the energy or Fourier spectra of aperiodic substitutional sequences.

The first connections between the irreducible Pisot property and the Fourier spectrum of a substitutional sequence were reported in Refs. [86,87], where it was conjectured that if the Perron–Frobenius eigenvalue of a substitutional system is a Pisot–Vijayaraghavan number, then the system is quasiperiodic. Later studies have revealed more details, providing a more sophisticated classification of substitutional systems with respect to the nature of their diffraction spectrum. In the one-dimensional case, sequences produced from irreducible Pisot substitutions have pure point Fourier spectra [88]. (I) The Pisot property, together with (II) the extra condition $\lambda \neq 0$, provide the means to distinguish between:

(1) *strictly quasiperiodic* sequences, in which the rank of the reciprocal lattice n_r is finite and larger than the dimension of the physical space of the sequence m, and
(2) *limit-quasiperiodic* sequences, in which the rank of reciprocal lattice n_r is countably infinite (in a 1–1 correspondence with the natural numbers or integers).

The distinction criterion between categories (1) and (2) is the value of the determinant of **S**: unimodular **S** implies strict quasiperiodicity, otherwise the structure is limit-quasiperiodic [89–91]. Limit-quasiperiodic structures can be interpreted as a superposition of an infinite number of strictly quasiperiodic structures. Examples of *strictly quasiperiodic* structures are the classical Fibonacci sequence [92] as well as all the precious means sequences [93] and the Fibonacci-class sequences [94] (cf. Table 2, where several substitutional sequences studied in the literature are listed, together with their substitution rules and matrices). *Limit-quasiperiodic* structure representatives are the mixed means sequences with $n \geq m$ [95].

For substitutions not satisfying the above-mentioned conditions (I) and (II), the situation is more complex. In such cases, the Fourier spectrum can be:

(3) *limit-periodic*, i.e., a superposition of countably infinite periodic structures. Some examples are the period doubling sequence and metallic means sequences with $n = l(l+1)$ [96],
(4) *singular continuous*, i.e., non-constant, non-decreasing, continuous and has zero derivative, everywhere that the derivative exists. Examples are the Thue–Morse sequence [97–99] and metallic means sequences with $n \neq l(l+1)$ [96], or even
(5) *absolutely continuous*, such as the Rudin–Shapiro sequence [100,101].

Apart from the above-mentioned sequences, there are others for which the substitution is not primitive or the matrix **S** cannot even be defined at all. Examples of non-primitive substitutions include the sequences inspired by the Cantor set [102], maybe the most well-known deterministic fractal. A sequence for which a substitution matrix cannot be defined is the classical Kolakoski$(1,2)$ sequence [103,104], and generally Kolakoski(p,q) sequences where p is odd and q even or vice versa [105]. The situation is different when p and q are both even or odd; then, a primitive **S** can be defined. In the former case, the sequences have been classified as limit-periodic [106]. In the latter case, the irreducible Pisot property holds when $2(p+q) \geq (p-q)^2$, and **S** is also unimodular when $p = q \pm 2$ [105].

Table 2. Substitutional sequences studied in the literature, together with the alphabets through which they are defined, the corresponding substitution rules, and the substitution matrices. In the last row, the subscripts o and e in the substitution rules denote substitutions that are applied on odd and even positions in the sequence, respectively.

Sequence	$\mathcal{A}$	Substitution Rule	S
Fibonacci	{A, B}	$s(A) = AB \quad s(B) = A$	$\begin{pmatrix} 1 & 1 \\ 1 & 0 \end{pmatrix}$
Precious means	{A, B}	$s(A) = A^nB \quad s(B) = A$	$\begin{pmatrix} n & 1 \\ 1 & 0 \end{pmatrix}$
Fibonacci-class	{A, B}	$s(A) = B^{n-1}AB \quad s(B) = B^{n-1}A$	$\begin{pmatrix} 1 & 1 \\ n & n-1 \end{pmatrix}$
Mixed means	{A, B}	$s(A) = A^nB^m \quad s(B) = A$	$\begin{pmatrix} n & 1 \\ m & 0 \end{pmatrix}$
Metallic means	{A, B}	$s(A) = AB^n \quad s(B) = A$	$\begin{pmatrix} 1 & 1 \\ n & 0 \end{pmatrix}$
Period doubling	{A, B}	$s(A) = AB \quad s(B) = AA$	$\begin{pmatrix} 1 & 2 \\ 1 & 0 \end{pmatrix}$
Thue–Morse	{A, B}	$s(A) = AB \quad s(B) = BA$	$\begin{pmatrix} 1 & 2 \\ 1 & 0 \end{pmatrix}$
Rudin–Shapiro	{A, B, C, D}	$s(A) = AB \quad s(B) = AC$ $s(C) = DB \quad s(D) = DC$	$\begin{pmatrix} 1 & 1 & 0 & 0 \\ 1 & 0 & 1 & 0 \\ 0 & 1 & 0 & 1 \\ 0 & 0 & 1 & 1 \end{pmatrix}$
Triadic Cantor set	{A, B}	$s(A) = ABA \quad s(B) = BBB$	$\begin{pmatrix} 2 & 0 \\ 1 & 3 \end{pmatrix}$
Asymmetric Cantor set	{A, B}	$s(A) = ABAA \quad s(B) = BBBB$	$\begin{pmatrix} 3 & 0 \\ 1 & 4 \end{pmatrix}$
Generalized Cantor set (t, d)	{A, B}	$s(A) = A^{\frac{t-d}{2}}B^dA^{\frac{t-d}{2}} \quad s(B) = B^t$	$\begin{pmatrix} t-d & 0 \\ d & t \end{pmatrix}$
Kolakoski $(p = 2m, q = 2n)$	{A = pp, B = qq}	$s(A) = A^mB^m \quad s(B) = A^nB^n$	$\begin{pmatrix} m & n \\ m & n \end{pmatrix}$
Kolakoski $(p = 2m + 1, q = 2n + 1)$	{A = pp, B = pq, C = qq}	$s(A) = A^mBC^m \quad s(B) = A^mBC^n$ $s(C) = A^nBC^n$	$\begin{pmatrix} m & m & n \\ 1 & 1 & 1 \\ m & n & n \end{pmatrix}$
Kolakoski $(p = 2m, q = 2m + 1)$ or $(p = 2m + 1, q = 2m)$	{p, q}	$s_o(q) = p^q \quad s_o(p) = p^p$ $s_e(q) = q^q \quad s_e(p) = q^p$	undefinable

4. Energy Structure of Nucleic Acid Wires

The energy structure of a physical system is closely connected to many of its properties (electrical, magnetic, thermal, optical, et cetera). A useful –and closely related to experimental data—quantity that describes the energy structure of a given system is the density of states (DOS), which shows the number of states that can be occupied by electrons at each energy. It can be formally defined as

$$g(E) = \sum_k \delta(E - E_k), \quad (6)$$

where no spin degeneracies are included. The sum runs over all allowed states, each of which has an eigenenergy E_k. A closely related quantity is the integrated density of states (IDOS), defined as

$$IDOS(E) = \int_{-\infty}^{E} g(E')dE', \quad (7)$$

i.e., it is the number of states that have energy smaller than E. Discontinuities in the IDOS indicate the presence of energy gaps, and the height of an IDOS step gives information about the level population. For periodic systems, the regions of allowed energies lead to smooth parts in DOS or IDOS curves, separated by well defined gaps at specific energies, thus reflecting the continuous electronic band structure of a periodic crystal. On the contrary, the DOS and IDOS of random systems are rough, indicative of the presence of a multitude of gaps between the allowed energy levels. As it has to do with deterministic aperiodic sequences with a substitution rule, which reflects their self-similarity, it has been conjectured (and proven, in several specific cases) that their energy spectrum is singular continuous, i.e., in the thermodynamic limit, it exhibits an infinity of gaps and vanishing bandwidths [107].

Furthermore, for primitive substitutions described by a Hamiltonian corresponding to the WM, the following gap-labeling theorem has been introduced by Bellissard et al. [108]:

Theorem 5.13 of Ref [108]. *Let $\hat{H}$ be a Hamiltonian corresponding to the WM, where the coefficients (i.e., parameters) are determined by a primitive substitution on a finite alphabet. Then, the values of the IDOS of $\hat{H}$ on the spectral gaps in $[0,1]$ belong to the $\mathbb{Z}(\lambda_{PF}^{-1})$ module generated by the components of the eigenvectors $\vec{v}_{PF}$ and $\vec{v}_{PF,2}$ of the substitution matrices $\mathbf{S}$ and $\mathbf{S}_2$, respectively.*

From the above theorem, it follows that, in order to obtain the position of the gaps in the (normalized) IDOS of a primitive substitutional sequence within the WM, it is sufficient to know the substitution matrices of its legal 1- and 2-letter words (c.f. Section 3.3). Specifically, the gaps can be labeled by the negative powers of λ_{PF} times integral linear combinations of the components of $\vec{v}_{PF}$ and $\vec{v}_{PF,2}$ that lie within the interval $[0,1]$ [108,109]. For example, in the case of Fibonacci sequences, from the diagonalization of $\mathbf{S}$ (cf. Table 2), we get $\lambda_{PF} = \phi$ and $\vec{v}_{PF} = [\phi^{-1} \quad \phi^{-2}]^T$ (where ϕ is the golden ratio). Hence, the sequence consists of $\approx$61.8% A letters and $\approx$38.2% B letters. The legal 2-letter words in the Fibonacci sequence are BA, AB, and AA (i.e., BB is forbidden), thus the induced 2-substitution reads (cf. Section 3.3) $s_2(AA) = (AB)(BA)$, $s_2(AB) = (AB)(BA)$, $s_2(BA) = (AA)$, leading to the induced substitution matrix

$$\mathbf{S}_2 = \begin{pmatrix} 0 & 0 & 1 \\ 1 & 1 & 0 \\ 1 & 1 & 0 \end{pmatrix}. \tag{8}$$

The Perron–Frobenius eigenvector (cf. Section 3.2) of $\mathbf{S}_2$ is $\vec{v}_{PF,2} = [\phi^{-3} \quad \phi^{-2} \quad \phi^{-2}]^T$. Hence, the gaps can be labeled by integer linear combinations of negative powers of ϕ. Since every positive power of ϕ can be reduced to a linear expression of the form $\phi^g = N_g \phi + N_{g-1}$, where N_g is the Fibonacci number of generation g, and it also holds that $\phi^g + \phi^{-g} \in \mathbb{N}^*$, the situation can be reduced to an integral linear combination of 1 and ϕ. Thus, the positions of the gaps in the IDOS of a Fibonacci sequence within the WM can by given by

$$\{\mathcal{G}_n\} = \{n\phi \mod 1, \forall n \in \mathbb{Z}\}. \tag{9}$$

Another interesting remark, arising from the DOS values of a single-stranded Fibonacci DNA sequence consisting of G and C, is that the ratio among the distances between DOS of consecutive generations tends to ϕ [110]. The IDOS of periodic, several aperiodic, and random binary DNA sequences with G and A on the same strand, calculated within the WM, taking into account both diagonal and off-diagonal disorder is presented in Figure 3 [83]. Periodic sequences display two well defined bands, separated by a single energy gap (the largest among all cases). Thue–Morse, Fibonacci, Rudin–Shapiro, and Kolakoski sequences possesss a staircase-like IDOS, while the shape of random sequence IDOS resembles, albeit it is more disrupted, to that of Rudin–Shapiro, and its main energy gap is the smallest among all cases. The fractal, Cantor set based, sequences have a very rough spectrum. For all sequences, the value of the IDOS at the largest energy gap is equal to the occurrence percentage of A. Furthermore, it has been observed that there are steps in the IDOS, the relative value of which is equal to the occurrence percentages of the possible base-pair triplets [83]. This remark holds for all categories of deterministic aperiodic sequences, either generated by a primitive substitution matrix or not, such as Kolakoski (1,2), further connecting the specific base-pair sequence of a DNA segment with its energy structure. The above-mentioned IDOS steps and the corresponding values are marked (where possible) in the left vertical axes of Figure 3.

Apart from the sequential disorder, mentioned above, other disorder types are present or can be induced in nucleic acid sequences. In Ref. [111], the authors study single poly(CG) or poly(CT) DNA strands with diluted base-pairing, i.e., for example, the G sites are randomly attached to their complementary C sites, with a probability p. The C-G base pairs are renormalized onto the first strand, leading to two inter-penetrating lattices: a periodic one containing the G or T sites and a random one containing bare and renormalized C sites. The DOS for three indicative cases, i.e., for $p = 0, 0.5$, and 1, is presented at the top panels of Figure 4. When $p = 0$, there are two well defined bands arising from the periodicity of the segment. The band character is maintained for $p = 1$, with a smaller gap for poly(CG), while for poly(CT) the number of bands changes to three, reflecting the total number of different sites (since the renormalization procedure takes into account the original structure). When $p = 0.5$, for poly(CG), fluctuations of the same magnitude in both allowed energy regions arise and the singularities are rounded off due to the induced disorder. For poly(CT), the bands collapse at a single energy region, stronger fluctuations are present at smaller energies than at larger ones, and there is a persisting van Hove singularity exactly at the on-site energy of T. Hence, in this case, diluted base-pairing produces a gapless structure and keeps a number of states extended (around the on-site energy of T), which is an ideal scenario for charge transport.

Several human diseases are associated with aberrant DNA methylation, which is heritable during cell division but does not alter the DNA sequence. In Ref. [112], a poly(CG) single-stranded segment is considered, with methyl groups randomly connected with the 5-carbon of C bases (forming the so-called 5-methylcytosine), again, with probability p. For completely unmethylated or methylated segments, the DOS consists of two smooth bands, derived by the on-site energy of G and C, for $p = 0$, and of G and 5-methylcytosine for $p = 1$, respectively; the only difference is in the energy intervals of the allowed states. For $0 < p < 1$, the smooth profile of the DOS is degraded, since the presence of randomly distributed methyl groups along the chain introduces a small disorder, which in turn leads to an enhancement in the effective resistance that can reach one order of magnitude.

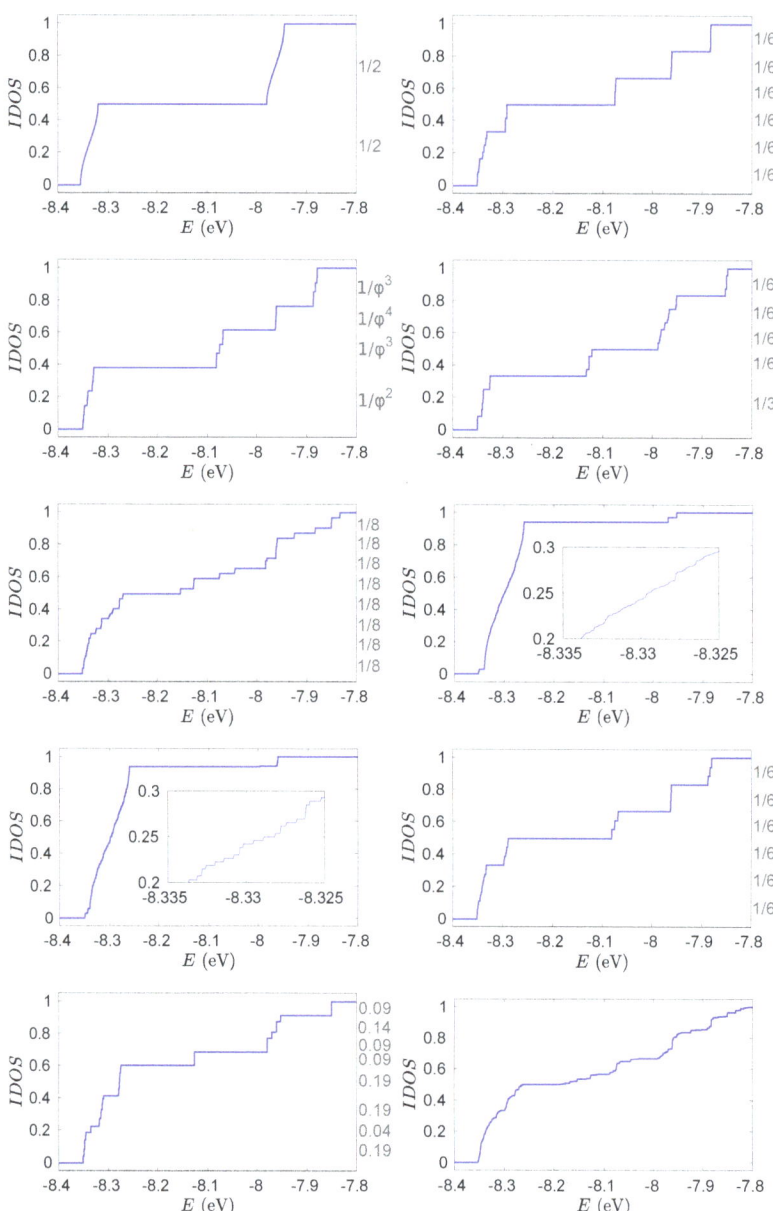

Figure 3. Normalized IDOS of various categories of binary DNA segments with purines on the same strand, within the WM. (**a**) Poly(GA); (**b**) Thue–Morse; (**c**) Fibonacci (**d**) Period-doubling; (**e**) Rudin–Shapiro; (**f**) Cantor Set; (**g**) Generalized Cantor Set (4,2); (**h**) Kolakoski(1,2); (**i**) Kolakoski(1,3); (**j**) Random (50% G, 50% A). Reprinted figure from K. Lambropoulos and C. Simserides, Periodic, quasiperiodic, fractal, Kolakoski, and random binary polymers: Energy structure and carrier transport, Phys. Rev. E **2019**, *99*, 032415 [83] http://dx.doi.org/10.1103/PhysRevE.99.032415, © 2019 by the American Physical Society.

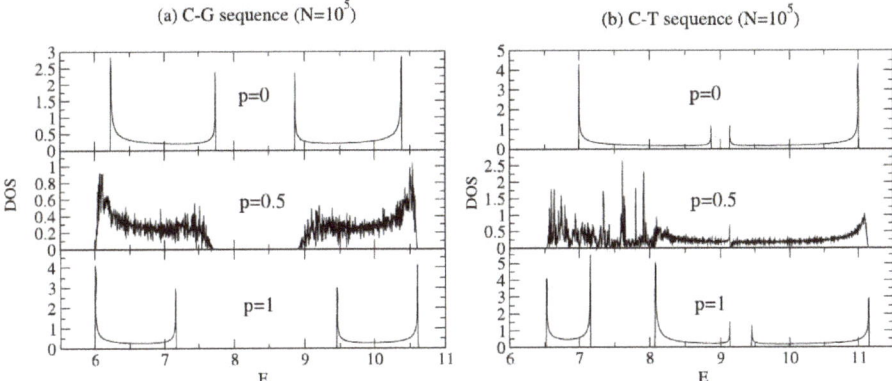

Figure 4. DOS of (**a**) poly(CG), and (**b**) poly(CT) DNA strands with diluted base-pairing at random cytosine sites with probability p. Figure reproduced with permission from F. A. B. F. de Moura, M. L. Lyra and E. L. Albuquerque, Electronic transport in poly(CG) and poly(CT) DNA segments with diluted base pairing, J. Phys. Condens. Matter **2008**, *20*, 075109 [111] http://dx.doi.org/10.1088/0953-8984/20/7/075109, © 2008 IOP Publishing. All rights reserved.

5. Coupling Nucleic Acids with Leads: Transmission Coefficients

In order to study the charge transport properties of nucleic acid nanowires within a TB framework, the system under examination is attached to two semi-infinite homogeneous metallic leads, which play the role of a carrier bath. The leads are characterized by a single on-site energy, ϵ_M, and a single hopping integral, t_M, so that the allowed energy states of the incident and outgoing waves lie in the interval $[\epsilon_M - 2|t_M|, \epsilon_M + 2|t_M|]$. Since detailed information on the nucleic acid's chemical bonding at the contacts is not known, one introduces effective parameters dealing with the tunneling probability between the frontier orbitals, roughly encompassing the bonding effects at the interface [113]. These parameters are $t_{R(L)}$ and couple the left (right) lead with the nucleic acid wire.

A first useful physical quantity to evaluate the charge transport properties of a quantum system is the transmission coefficient, $T(E)$. It is an energy-dependent quantity that describes the probability that a carrier, incident to a quantum wire, transmits through its eigenstates. Charge transport will experience a sequence-dependent contribution of backscattering, according to the distribution of potential barriers, corresponding to bases or base pairs, over the length scale of the sequence [24].

The coupling between the nucleic acid and the leads plays an important role on the transmission profiles. It has been shown that *stronger coupling does not necessarily mean higher transmission*. In Ref. [114], the authors studied the transmission profiles of a single-stranded poly(GACT) DNA chain within the WM, with purely diagonal disorder, assuming equal coupling parameters with both leads ($t_R = t_{cL} = \tau$), and arrived at the resonance condition $\tau = \sqrt{t_M t}$, where t is the hopping integral between the wire sites. From Figure 5, it is evident that, when the value of the coupling parameter is either smaller or larger than the one fulfilling the resonance condition, quite smaller transmission peaks are obtained. This result properly illustrates the influence of contacts on electrical transport. This extreme sensitivity is due to interference effects between the DNA molecular bands and the electronic structure of the leads at the lead-DNA interface.

The above-mentioned results were generalized in an analytical manner for any periodic WM, through the conditions $\omega = \frac{t_M t_u}{t_R t_L} = \pm 1$ (ideal coupling condition), where t_u couples the moieties at the end of a unit cell and at the start of the next, and $\chi = \frac{t_L}{t_R} = \pm 1$ (symmetric coupling condition) [60]. The ideal coupling condition, $\omega = \pm 1$, implies that the system and the leads are interconnected as if they were connected to themselves. When this condition is reached, the existence of fully resonant states is guaranteed at specific energies determined by the zeros of Chebyshev polynomials of the second kind [115]. Hence, any periodic sequence can display full transmission, if appropriate couplings are utilized. Deviations from the symmetric coupling condition give rise to secondary peaks. The effect of the coupling strength and the asymmetry factors, together with the internal hoppings, is exemplified in Figure 6, for a generic periodic WM with two sites per unit cell (hence, two hopping integrals t_1 and t_2 connect the wire sites) and $N = 10$. It is evident that the ideal and symmetric coupling conditions lead to the most efficient transmission. For ideal and asymmetric coupling, except for the peaks of magnitude 1, there is one additional peak, which is of significant magnitude only when $|t_1| \approx |t_2|$. In the strong (weak) and symmetric coupling regimes, the peaks that are closer to the band gap vanish (emerge) as $\left|\frac{t_1}{t_2}\right|$ increases. When the coupling is asymmetric, transmission is enhanced only in one of the two bands.

Analogous conclusions can be obtained for more complex TB models. In Ref. [31], a poly(G)-poly(C) oligomer ($N = 5$) was studied within the FM. The authors report that, for small values of coupling, the transmission shows sharp and narrow unit resonances due to the localization of states, while, as the coupling increases, the well-arranged resonant peaks overlap. An inspection of Figure 7 of Ref. [31] indicates that there are intermediate values of $t_L(=t_R)$ in which the overall transmission is more enhanced compared to smaller and larger values.

In Ref. [116], the authors study a poly(G)-poly(C) chain within an extension of the FLM, which allows hopping between backbone sites as well as all possible diagonal hoppings (between the nitrogenous bases as well as between the bases and the backbone). Each of the two strands containing the DNA bases is connected with each lead with equal coupling parameters. For diagonal hoppings being switched either on or off, it can again be concluded that stronger coupling with the leads does not necessarily lead to enhanced transmission (cf. the panels in the first two rows in Figure 7). This is also evident by comparing the averaged transmission coefficient, which is defined as

$$T_a(E) = \frac{\int_{E_{\min}}^{E} T(e) de}{E - E_{\min}}, \qquad (10)$$

cf. the panels in third row of Figure 7. Although $T(E)$ and $T_a(E)$ are indeed much smaller for $t_L = t_R = 0.1$ eV, an increase from 0.5 eV to 0.9 eV does not lead to transmission enhancement. In fact, for diagonal hoppings switched both on and off, T_a reaches larger values for the intermediate coupling $t_L = t_R = 0.5$ eV.

The above discussion demonstrates that, apart from the internal degree of disorder of a given sequence, other factors can significantly affect their charge transport properties.

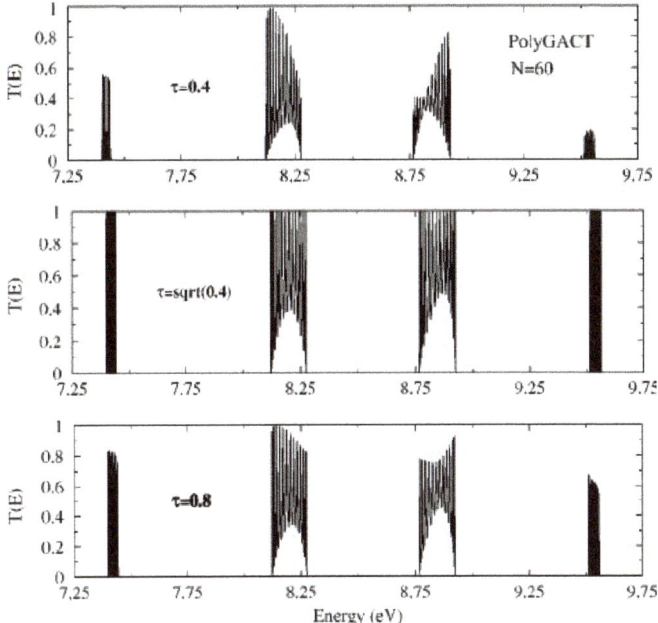

Figure 5. Transmission coefficient for a poly(GACT) chain within the WM, with $N = 60$, $t_M = 1.0$ eV, $t = 0.4$ eV, and $\tau = 0.4$ eV; i.e., $\tau = \sqrt{t_M t}$ (**top**), $\tau = \sqrt{0.4}$ eV (**middle**), $\tau = \sqrt{0.8}$ eV (**bottom**). Reprinted figure with permission from E. Maciá, F. Triozon, and S. Roche, Contact-dependent effects and tunneling currents in DNA molecules, Phys. Rev. B **2005**, *71*, 113106 [114] http://dx.doi.org/10.1103/PhysRevB.71.113106, © 2005 by the American Physical Society.

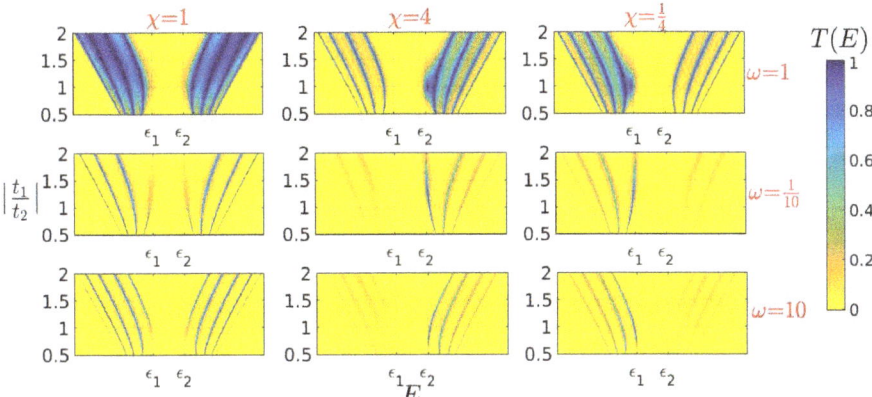

Figure 6. Transmission coefficient of a periodic WM with two sites per unit cell and $N = 10$ for ideal (**top**), strong (**middle**), and weak (**bottom**) coupling with the leads. (Left column) Symmetric coupling. (Middle column) Asymmetric coupling with $|\chi| > 1$. (Right column) Asymmetric coupling with $|\chi| < 1$. The leads parameters are such that all the eigenstates of the system are contained. Reprinted from Ref. [60], K. Lambropoulos and C Simserides, Spectral and transmission properties of periodic 1D tight-binding lattices with a generic unit cell: an analysis within the transfer matrix approach, J. Phys. Commun. **2018**, *2*, 035013 [60] http://dx.doi.org/10.1088/2399-6528/aab065, CC BY 3.0.

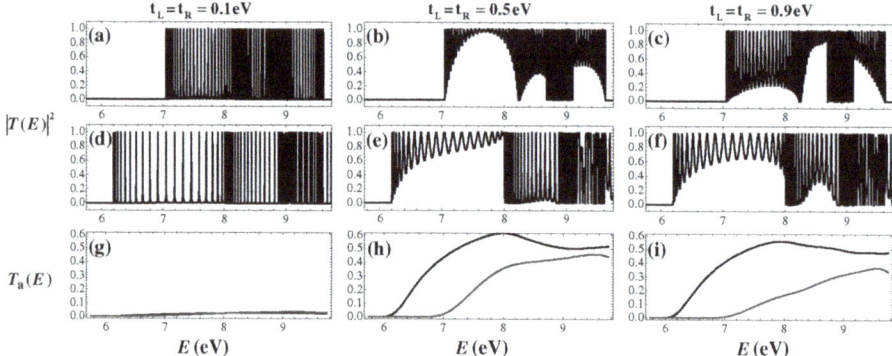

Figure 7. Transmission spectra as a function of energy without (**a**–**c**) and with (**d**–**f**) the diagonal hoppings; (**g**–**i**) average transmission spectra as a function of energy: gray line (diagonal hoppings switched off) and black line (diagonal hoppings switched on). (Left column) $t_L = t_R = 0.1$ eV. (Middle column) $t_L = t_R = 0.5$ eV. (Right column) $t_L = t_R = 0.9$ eV. Reprinted from S. Malakooti, E. R. Hedin, Y. D. Kim, and Y. S. Joe, Enhancement of charge transport in DNA molecules induced by the next nearest-neighbor effects, J. Appl. Phys. **2012**, *112*, 094703 [116], http://dx.doi.org/10.1063/1.4764310, with the permission of AIP Publishing.

6. Current–Voltage Curves

The situation is more complex as far as the calculation of $I - V$ characteristic curves is concerned. The $I - V$ curve of a given nucleic acid segment can be given, using the Landauer–Büttiker formalism [61,117,118], by the relation

$$I(V) = \frac{2e}{h} \int_{-\infty}^{\infty} T(E,V)[f_L(E - \mu_L) - f_R(E - \mu_R)]dE, \qquad (11)$$

under the assumption that charge propagates from left to right. $\mu_{L(R)}$ and $f_{L(R)}(E)$ are the chemical potential and the Fermi–Dirac distribution at the left (right) lead, respectively. From Equation (11), we deduce that there are several factors, apart from the structure of the sequence under examination that have an effect on the magnitude of currents, the bias regime and the shape of the $I - V$ curves. These factors include:

(a) The choice of the Fermi level of the leads E_F, which coincides with ϵ_M if one electron per site is assumed. If E_F is not aligned with an allowed energy region of the segment, then no currents occur in the vicinity of $V = 0$, while a metallic behavior is expected otherwise.

(b) The way the external bias is applied. For example, only one of the leads' energy bands can be shifted, so that $\mu_L = E_F + eV$, and $\mu_R = E_F$, or, alternatively, both leads' bands can be symmetrically shifted so that $\mu_{\frac{L}{R}} = E_F \pm \frac{eV}{2}$. This choice affects both the way the voltage drop is induced in the nucleic acid sequence and the energy limits of the conductance channel. At zero temperature, the Fermi–Dirac distributions become Heaviside step functions and determine the limits of integration. Hence, Equation (11) can be simplified to

$$I(V) = \frac{2e}{h} \int_{\mu_R}^{\mu_L} T(E,V)dE, \qquad (12)$$

while, at finite temperatures, it can be written in the form

$$I(V) = \frac{2e}{h} \sinh\left(\frac{eV}{2k_B T}\right) \int_{-\infty}^{\infty} \frac{T(E,V)dE}{\cosh\left(\frac{E-E_F}{k_B T}\right) + \cosh\left(\frac{eV}{2k_B T}\right)}, \qquad (13)$$

i.e., the $I-V$ curve occurs from the modulation of the hyperbolic function $\sinh\left(\frac{eV}{2k_BT}\right)$ by the integral factor expression [119].

(c) Whether or not the transmission coefficient is considered as bias-dependent. Although assuming bias-independent transmission coefficient could be a justified choice in the small bias regime, and it is indeed less computationally costly, this assumption cannot lead, under any circumstances, to the occurrence of negative differential resistance, since an increasingly larger part (as V increases) of a nonnegative function is integrated.

There are several works discussing the $I-V$ curves of nucleic acid sequences, considering different types of order or disorder. Regarding sequential order or disorder, in Ref. [83], the $I-V$ curves of periodic, deterministic aperiodic, and random binary DNA segments have been studied within the WM. The curves have been shown to have clearly distinct shapes for different sequence categories. It has also been demonstrated that periodic sequences lead to the most enhanced currents. Additionally, there are several categories deterministic aperiodic sequences (specifically, Fibonacci, Period-doubling, Cantor and generalized Cantor) that can also display significant currents, depending on the Fermi level of the leads. Random sequences represent the least efficient category, since they were found to always display smaller currents than all their deterministic aperiodic counterparts with similar base-pair content.

In Ref. [120], the authors study dry and hydrated DNA sequences with correlated and uncorrelated disorder within a WM, for $N=50$ and at a temperature of 300 K. For different concentrations of G and A sites, the resulting currents are larger for correlated disorder, both for dry and backbone-hydrated sequences. Generally, the authors report a conductor to semiconductor to insulator transition as a function of three effects, i.e., sequence size, disorder, and hydration, suggesting that an appropriate choice of chain size and relative concentration of base pairs can be used to tailor the electrical behavior of DNA strands.

A similar transition has been reported by introducing conformal variation at the helical symmetry as well as backbone disorder into a FLM [121]. Helical symmetry is taken into account via the inclusion of hopping integrals between bases in adjacent pitches (i.e., turns of the helix). The number of base-pairs within a given pitch is denoted by n. Backbone disorder is introduced by a random distribution of backbone on-site energies, characterized by a disorder strength w. The results for poly(G)-poly(C) and poly(A)-poly(T) chains with $N=50$, for different values of n and w are shown in Figure 8. At low disorder, the effect of n is smaller, since, in that case, any path of charge conduction is equivalent, as an electron feels almost no potential variation. As the disorder increases, the effect of n becomes more distinctive, since there is substantial variation of the effective potential at different sites and an increase of n gives an electron more shortcut pathways to move along the DNA chain. The current is enhanced with increasing n, and the effect is more vivid for strong disorder. Furthermore, for weak disorder, a cut-off voltage is observed in the $I-V$ curves, which reduces with increasing n. At strong disorder, the current is enhanced and almost linear response is observed at larger values of n, which indicates a transition from the insulating to the metallic phase.

Thermal structural disorder has been studied in Ref. [122], by introducing a random variation in the hopping integrals of a poly(G)-poly(C) chain with $N=5$, within an FLM allowing inter-backbone hoppings. Comparing the $I-V$ curves of such systems for $T=0$ K and $T=300$ K, the authors report that the voltage threshold for current onset is about the same, indicating that the thermal structural disorder does not affect the voltage gaps. Above that threshold, as the temperature increases, the linear behavior of the current changes to a step-like behavior, and the current is reduced, since the static distortion increases elastic scattering of electrons through the DNA molecule.

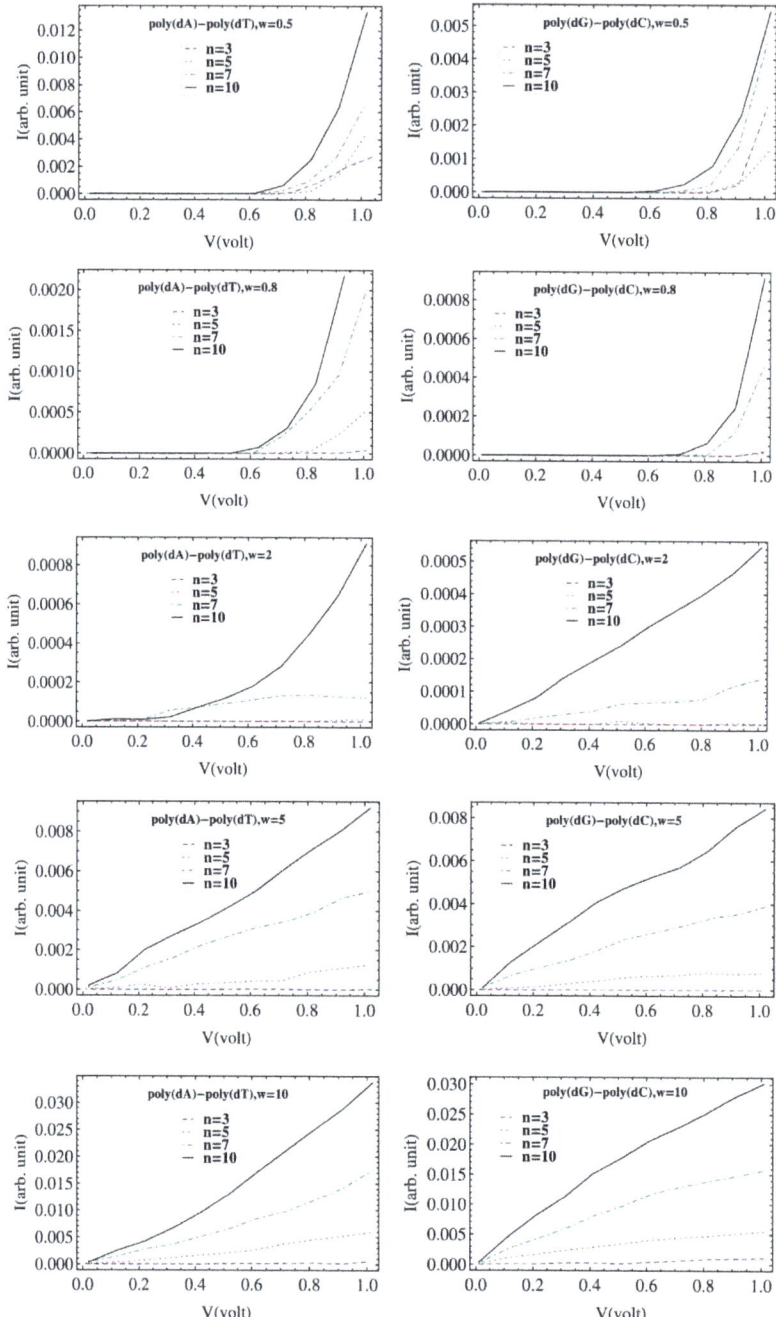

Figure 8. $I - V$ curves for poly(dA)-poly(dT) and poly(dG)-poly(dC) various disorder strengths w and pitch-size values n. For weak disorder, the cut-off voltage reduces with n, showing semiconducting behaviour. For strong disorder, the current is considerably enhanced with increasing n, giving a insulator to metal transition. Reproduced from Ref. [121], S. Kundu and S. N. Karmakar, Conformation dependent electronic transport in a DNA double-helix, AIP Adv. **2015**, 5, 107122 [121] http://dx.doi.org/10.1063/1.4934507, CC BY 3.0.

The effect of cytosine methylation disorder on the $I-V$ curves of a single stranded GAGCTGACGTTCACGG segment retrieved from the first sequenced human chromosome (chromosome 22) has been studied within the WM in Ref. [123]. The effect of all possible single, double, and triple methylation defects (out of the totally four C sites) is addressed. It is demonstrated that even a single methylated site reduces the currents by one order of magnitude. This reduction is directly associated with the fact that such sites act as additional impurity centers. The observed sensitivity of the saturation current on the position of the methylated cytosine is related to the impact of methylation on the hopping integrals to the neighboring bases. Thus, for a single methylation defect, the saturation current is strongly suppressed when cytosine is connected with guanine; for two defects, this suppression is smaller; for three methylations, the non-methylated base is the one that acts as a defect, hence the suppression of the saturation current will be larger when the cytosine has both hopping amplitudes to the neighboring bases enhanced by methylation. These results suggest the feasibility of using $I-V$ curves to develop biosensors for the purpose of diagnosis.

There also exist efforts aiming to examine the potentiality to utilize the charge transport characteristics of nucleic acids as a tool to probe several diseases or disorders. In Ref. [124], the $I-V$ characteristics of twenty seven single-stranded microRNA chains (with 21 to 23 nucleotides) related to the autism spectrum disorder have been studied. The authors classified the chains into five groups according to their conductivity (from high to negligible), suggesting that a kind of electronic biosensor can be developed to distinguish different profiles of autism disorders.

In Ref. [125], a similar treatment was employed to study DNA sequences related to the Huntington's Disease. A segment of the human chromosome 4p16.3 was modified by the addition of a variant number of CAG repeats, the number of which determines whether a person does or does not have Huntington's Disease; repeats smaller than 27 are normal; repeats between 27 and 35 are rarely associated with the disease, but it may expand in paternal transmission; repeats between 36 and 39 are associated with reduced penetration, so individuals may or may not develop the disease; 40 and above are associated with the disease [126]. The increasing presence of periodicity leads to enhanced transmission and thus to more efficient electronic transport. $I-V$ calculations revealed that the above-mentioned groups based on the number of repeats can be characterized by different value ranges for the saturation currents, indicating a promising method for identifying Huntington's disease.

7. Conclusions

This review was devoted to tight-binding (TB) modeling of nucleic acid sequences like DNA and RNA. We briefly presented the TB approach and discussed its various submodels: wire, ladder, extended ladder, fishbone (wire), and fishbone ladder. We addressed various types of orders (periodic, quasiperiodic, fractal) or disorder (diagonal, non-diagonal, random, methylation) and explained how these various types of order or disorder affect charge transport. We proceeded to a discussion of aperiodicity, quasicrystals and the mathematics of aperiodic substitutional sequences. Specifically, we discussed the notions of primitive substitutions, Perron–Frobenius eigenvalue, induced substitutions, and Pisot property. We explained how the energy structure of nucleic acid wires is affected by order or disorder. We also discussed the corresponding transmission coefficients, focusing on the effects of coupling the nucleic acids to external leads, and demonstrating that, apart from the internal degree of order or disorder of a given sequence, there are several other factors that can significantly affect their charge transport properties. We also discussed the effects that various types of order or disorder induce on the current–voltage curves and presented some efforts aiming to examine the potentiality to utilize the charge transport characteristics of nucleic acids as a tool to probe several diseases or disorders. The sensitivity that the results demonstrate regarding the choice of the nucleic acids sequence, the recruited models and parametrizations, the way the systems are coupled to external leads, the nature of the leads, the environmental conditions, etc, indicate that much work is needed in order to reach a thorough description of the effect the combination such a multitude of factors has on charge transport. Furthermore, other factors, such as the sequence geometry or the use of modified nitrogenous bases could be potentially used to tailor the above-mentioned properties of nucleic acids and analogues.

Author Contributions: Conceptualization, K.L. and C.S.; methodology, K.L. and C.S.; software, K.L. and C.S.; validation, K.L. and C.S.; formal analysis, K.L. and C.S.; investigation, K.L. and C.S.; resources, K.L. and C.S.; data curation, K.L. and C.S.; writing—original draft preparation, K.L.; writing—review and editing, C.S.; visualization, K.L. and C.S.; supervision, C.S.; project administration, C.S.; funding acquisition, K.L. and C.S.

Funding: This research was surpported by the Hellenic Foundation for Research and Innovation (HFRI) and the General Secretariat for Research and Technology (GSRT), under the HFRI PhD Fellowship grant (GA no 260).

Conflicts of Interest: The authors declare no conflict of interest.

Abbreviations

The following abbreviations are used in this manuscript:

DNA	Deoxyribonucleic Acid
RNA	Ribonucleic acid
G	Guanine
A	Adenine
C	Cytosine
T	Thymine
U	Uracil
TB	Tight-Binding
WM	Wire Model
LM	Ladder Model
ELM	Extended Ladder Model
FM	Fishbone Model
FLM	Fishbone Ladder Model
DOS	Density of States
IDOS	Integrated Density of States

References

1. Watson, J.D.; Crick, F.H.C. Molecular Structure of Nucleic Acids: A Structure for Deoxyribose Nucleic Acid. *Nature* **1953**, *171*, 737–738.10.1038/171737a0. [CrossRef] [PubMed]
2. Page, C.C.; Moser, C.C.; Dutton, P.L. Mechanism for electron transfer within and between proteins. *Curr. Opin. Chem. Biol.* **2003**, *7*, 551–556.10.1016/j.cbpa.2003.08.005. [CrossRef] [PubMed]
3. Gray, H.B.; Winkler, J.R. Electron flow through metalloproteins. *Biochim. Biophys. Acta* **2010**, *1797*, 1563–1572.10.1016/j.bbabio.2010.05.001. [CrossRef] [PubMed]
4. Moser, C.C.; Ross Anderson, J.L.; Dutton, P.L. Guidelines for tunneling in enzymes. *Biochim. Biophys. Acta* **2010**, *1797*, 1573–1586.10.1016/j.bbabio.2010.04.441. [CrossRef] [PubMed]
5. Artés, J.M.; López-Martínez, M.; Díez-Pérez, I.; Sanz, F.; Gorostiza, P. Nanoscale charge transfer in redox proteins and DNA: Towards biomolecular electronics. *Electrochim. Acta* **2014**, *140*, 83–95.10.1016/j.electacta.2014.05.089. [CrossRef]
6. Kannan, A.M.; Renugopalakrishnan, V.; Filipek, S.; Li, P.; Audette, G.F.; Munukutla, L. Bio-Batteries and Bio-Fuel Cells: Leveraging on Electronic Charge Transfer Proteins. *J. Nanosci. Nanotechnol.* **2009**, *9*, 1665–1678.10.1166/jnn.2009.si03. [CrossRef] [PubMed]
7. Dandliker, P.J.; Holmlin, R.E.; Barton, J.K. Oxidative Thymine Dimer Repair in the DNA Helix. *Science* **1997**, *275*, 1465–1468.10.1126/science.275.5305.1465. [CrossRef]
8. Rajski, S.R.; Jackson, B.A.; Barton, J.K. DNA repair: models for damage and mismatch recognition. *Mutat. Res.* **2000**, *447*, 49–72.10.1016/s0027-5107(99)00195-5. [CrossRef]
9. Giese, B. Electron transfer through DNA and peptides. *Bioorg. Med. Chem.* **2006**, *14*, 6139–6143.10.1016/j.bmc.2006.05.067. [CrossRef]
10. Shih, C.T.; Cheng, Y.Y.; Wells, S.A.; Hsu, C.L.; Römer, R.A. Charge transport in cancer-related genes and early carcinogenesis. *Comput. Phys. Commun.* **2011**, *182*, 36–38.10.1016/j.cpc.2010.06.029. [CrossRef]

11. Endres, R.G.; Cox, D.L.; Singh, R.R.P. Colloquium: The quest for high-conductance DNA. *Rev. Mod. Phys.* **2004**, *76*, 195–214.10.1103/revmodphys.76.195. [CrossRef]
12. Wohlgamuth, C.H.; McWilliams, M.A.; Slinker, J.D. DNA as a Molecular Wire: Distance and Sequence Dependence. *Anal. Chem.* **2013**, *85*, 8634–8640.10.1021/ac401229q. [CrossRef] [PubMed]
13. Abouzar, M.H.; Poghossian, A.; Cherstvy, A.G.; Pedraza, A.M.; Ingebrandt, S.; Schöning, M.J. Label-free electrical detection of DNA by means of field-effect nanoplate capacitors: Experiments and modeling. *Phys. Status Solidi A* **2012**, *209*, 925–934.10.1002/pssa.201100710. [CrossRef]
14. Ye, Y.J.; Shen, L.L. DFT approach to calculate electronic transfer through a segment of DNA double helix. *J. Comput. Chem.* **2000**, *21*, 1109–1117.10.1002/1096-987X(200009)21:12<1109::AID-JCC7>3.0.CO;2-4. [CrossRef]
15. Ye, Y.J.; Jiang, Y. Electronic structures and long-range electron transfer through DNA molecules. *Int. J. Quantum Chem.* **2000**, *78*, 112–130.10.1002/(SICI)1097-461X(2000)78:2<112::AID-QUA5>3.0.CO;2-5. [CrossRef]
16. Voityuk, A.A. Electronic couplings and on-site energies for hole transfer in DNA: Systematic quantum mechanical/molecular dynamic study. *J. Chem. Phys.* **2008**, *128*, 115101, doi:10.1063/1.2841421. [CrossRef]
17. Kubař, T.; Woiczikowski, P.B.; Cuniberti, G.; Elstner, M. Efficient Calculation of Charge-Transfer Matrix Elements for Hole Transfer in DNA. *J. Phys. Chem. B* **2008**, *112*, 7937–7947.10.1021/jp801486d. [CrossRef]
18. Tassi, M.; Morphis, A.; Lambropoulos, K.; Simserides, C. RT-TDDFT study of hole oscillations in B-DNA monomers and dimers. *Cogent Phys.* **2017**, *4*, 1361077, doi:10.1080/23311940.2017.1361077. [CrossRef]
19. Artacho, E.; Machado, M.; Sánchez-Portal, D.; Ordejón, P.; Soler, J.M. Electrons in dry DNA from density functional calculations. *Mol. Phys.* **2003**, *101*, 1587–1594.10.1080/0026897031000068587. [CrossRef]
20. Mehrez, H.; Anantram, M.P. Interbase electronic coupling for transport through DNA. *Phys. Rev. B* **2005**, *71*, 115405, doi:10.1103/PhysRevB.71.115405. [CrossRef]
21. Adessi, C.; Walch, S.; Anantram, M.P. Environment and structure influence on DNA conduction. *Phys. Rev. B* **2003**, *67*, 081405, doi:10.1103/PhysRevB.67.081405. [CrossRef]
22. Barnett, R.N.; Cleveland, C.L.; Landman, U.; Boone, E.; Kanvah, S.; Schuster, G.B. Effect of Base Sequence and Hydration on the Electronic and Hole Transport Properties of Duplex DNA: Theory and Experiment. *J. Phys. Chem. A* **2003**, *107*, 3525–3537.10.1021/jp022211r. [CrossRef]
23. Cuniberti, G.; Craco, L.; Porath, D.; Dekker, C. Backbone-induced semiconducting behavior in short DNA wires. *Phys. Rev. B* **2002**, *65*, 241314, doi:10.1103/PhysRevB.65.241314. [CrossRef]
24. Roche, S.; Bicout, D.; Maciá, E.; Kats, E. Long Range Correlations in DNA: Scaling Properties and Charge Transfer Efficiency. *Phys. Rev. Lett.* **2003**, *91*, 228101, doi:10.1103/PhysRevLett.91.228101. [CrossRef]
25. Roche, S. Sequence Dependent DNA-Mediated Conduction. *Phys. Rev. Lett.* **2003**, *91*, 108101, doi:10.1103/PhysRevLett.91.108101. [CrossRef]
26. Palmero, F.; Archilla, J.F.R.; Hennig, D.; Romero, F.R. Effect of base-pair inhomogeneities on charge transport along the DNA molecule, mediated by twist and radial polarons. *New J. Phys.* **2004**, *6*, 13 doi:10.1088/1367-2630/6/1/013. [CrossRef]
27. Yamada, H. Localization of electronic states in chain models based on real DNA sequence. *Phys. Lett. A* **2004**, *332*, 65–73.10.1016/j.physleta.2004.09.041. [CrossRef]
28. Apalkov, V.M.; Chakraborty, T. Electron dynamics in a DNA molecule. *Phys. Rev. B* **2005**, *71*, 033102, doi:10.1103/PhysRevB.71.033102. [CrossRef]
29. Klotsa, D.; Römer, R.A.; Turner, M.S. Electronic Transport in DNA. *Biophys. J.* **2005**, *89*, 2187–2198.10.1529/biophysj.105.064014. [CrossRef]
30. Shih, C.T.; Roche, S.; Römer, R.A. Point-Mutation Effects on Charge-Transport Properties of the Tumor-Suppressor Gene *p53*. *Phys. Rev. Lett.* **2008**, *100*, 018105, doi:10.1103/PhysRevLett.100.018105. [CrossRef]
31. Joe, Y.S.; Lee, S.H.; Hedin, E.R. Electron transport through asymmetric DNA molecules. *Phys. Lett. A* **2010**, *374*, 2367–2373.10.1016/j.physleta.2010.03.050. [CrossRef]
32. Yi, J. Conduction of DNA molecules: A charge-ladder model. *Phys. Rev. B* **2003**, *68*, 193103, doi:10.1103/PhysRevB.68.193103. [CrossRef]
33. Caetano, R.A.; Schulz, P.A. Sequencing-Independent Delocalization in a DNA-Like Double Chain with Base Pairing. *Phys. Rev. Lett.* **2005**, *95*, 126601, doi:10.1103/PhysRevLett.95.126601. [CrossRef]

34. Wang, X.F.; Chakraborty, T. Charge Transfer via a Two-Strand Superexchange Bridge in DNA. *Phys. Rev. Lett.* **2006**, *97*, 106602, doi:10.1103/PhysRevLett.97.106602. [CrossRef]
35. Ashcroft, N.W.; Mermin, N.D. *Solid State Physics*; Saunders College: Philadelphia, PA, USA, 1976.
36. Slater, J.C.; Koster, G.F. Simplified LCAO Method for the Periodic Potential Problem. *Phys. Rev.* **1954**, *94*, 1498–1524.10.1103/PhysRev.94.1498. [CrossRef]
37. Harrison, W.A. *Electronic Structure and the Properties of Solids: The Physics of the Chemical Bond (Dover Books on Physics)*, 2nd ed.; Dover Publications: Ney York, NY, USA, 1989.
38. Papaconstantopoulos, D.A.; Mehl, M.J. The Slater-Koster tight-binding method: A computationally efficient and accurate approach. *J. Phys. Condens. Matter* **2003**, *15*, 413–440, doi:10.1088/0953-8984/15/10/201. [CrossRef]
39. Foulkes, W.M.C. Tight-Binding Models and Coulomb Interactions for *s*, *p*, and *d* Electrons. In *Quantum Materials: Experiments and Theory*; Pavarini, E., Koch, E., van den Brink, J., Sawatzky, G., Eds.; Modeling and Simulation, Forschungszentrum Jülich: Jülich, Germany, 2016; Volume 6.
40. Molinari, L. Transfer matrices and tridiagonal-block Hamiltonians with periodic and scattering boundary conditions. *J. Phys. A Math. Gen.* **1997**, *30*, 983–997, doi:10.1088/0305-4470/30/3/021. [CrossRef]
41. Molinari, L. Spectral duality and distribution of exponents for transfer matrices of block-tridiagonal Hamiltonians. *J. Phys. A Math. Gen.* **2003**, *36*, 4081–4090.10.1088/0305-4470/36/14/311. [CrossRef]
42. Cuniberti, G.; Maciá, E.; Rodríguez, A.; Römer, R.A. Tight-Binding Modeling of Charge Migration in DNA Devices. In *Charge Migration in DNA: Perspectives from Physics, Chemistry, and Biology*; Chakraborty, T., Ed.; Springer: Berlin/Heidelberg, Germany, 2007; pp. 1–20, doi:10.1007/978-3-540-72494-0_1.
43. Albuquerque, E.L.; Fulco, U.L.; Freire, V.N.; Caetano, E.W.S.; Lyra, M.L.; de Moura, F.A.B.F. DNA-based nanobiostructured devices: The role of quasiperiodicity and correlation effects. *Phys. Rep.* **2014**, *535*, 139–209.10.1016/j.physrep.2013.10.004. [CrossRef]
44. Zarea, M.; Berlin, Y.; Ratner, M.A. Effect of the reflectional symmetry on the coherent hole transport across DNA hairpins. *J. Chem. Phys.* **2017**, *146*, 114105.10.1063/1.4978571. [CrossRef]
45. Simserides, C. A systematic study of electron or hole transfer along DNA dimers, trimers and polymers. *Chem. Phys.* **2014**, *440*, 31–41, doi:10.1016/j.chemphys.2014.05.024. [CrossRef]
46. Albuquerque, E.L.; Vasconcelos, M.S.; Lyra, M.L.; de Moura, F.A.B.F. Nucleotide correlations and electronic transport of DNA sequences. *Phys. Rev. E* **2005**, *71*, 021910.10.1103/PhysRevE.71.021910. [CrossRef]
47. Lambropoulos, K.; Chatzieleftheriou, M.; Morphis, A.; Kaklamanis, K.; Lopp, R.; Theodorakou, M.; Tassi, M.; Simserides, C. Electronic structure and carrier transfer in B-DNA monomer polymers and dimer polymers: Stationary and time-dependent aspects of a wire model versus an extended ladder model. *Phys. Rev. E* **2016**, *94*, 062403.10.1103/PhysRevE.94.062403. [CrossRef]
48. Xu, M.; Endres, R.; Tsukamoto, S.; Kitamura, M.; Ishida, S.; Arakawa, Y. Conformation and Local Environment Dependent Conductance of DNA Molecules. *Small* **2005**, *1*, 1168–1172.10.1002/smll.200500216. [CrossRef]
49. Iguchi, K. π-electrons in a single strand DNA: A phenomenological approach. *Int. J. Mod. Phys. B* **2004**, *18*, 1845–1910.10.1142/S0217979204025051. [CrossRef]
50. Kundu, S.; Karmakar, S.N. Localization phenomena in a DNA double-helix structure: A twisted ladder model. *Phys. Rev. E* **2014**, *89*, 032719.10.1103/PhysRevE.89.032719. [CrossRef]
51. Malakooti, S.; Hedin, E.; Joe, Y. Tight-binding approach to strain-dependent DNA electronics. *J. Appl. Phys.* **2013**, *114*, 014701.10.1063/1.4812394. [CrossRef]
52. Varela, S.; Mujica, V.; Medina, E. Effective spin–orbit couplings in an analytical tight-binding model of DNA: Spin filtering and chiral spin transport. *Phys. Rev. B* **2016**, *93*, 155436.10.1103/PhysRevB.93.155436. [CrossRef]
53. Díaz, E. Analysis of the interband optical transitions: Characterization of synthetic DNA band structure. *J. Chem. Phys.* **2008**, *128*, 175101.10.1063/1.2901046. [CrossRef]
54. Maciá, E. Electronic structure and transport properties of double-stranded Fibonacci DNA. *Phys. Rev. B* **2006**, *74*, 245105.10.1103/PhysRevB.74.245105. [CrossRef]
55. Maciá, E.; Roche, S. Backbone-induced effects in the charge transport efficiency of synthetic DNA molecules. *Nanotechnology* **2006**, *17*, 3002–3007.10.1088/0957-4484/17/12/031. [CrossRef]
56. Lambropoulos, K.; Chatzieleftheriou, M.; Morphis, A.; Kaklamanis, K.; Theodorakou, M.; Simserides, C. Unbiased charge oscillations in B-DNA: Monomer polymers and dimer polymers. *Phys. Rev. E* **2015**, *92*, 032725.10.1103/PhysRevE.92.032725. [CrossRef]

57. Lambropoulos, K.; Vantaraki, C.; Bilia, P.; Mantela, M.; Simserides, C. Periodic polymers with increasing repetition unit: Energy structure and carrier transfer. *Phys. Rev. E* **2018**, *98*, 032412.10.1103/PhysRevE.98.032412. [CrossRef]

58. MacKinnon, A.; Kramer, B. The scaling theory of electrons in disordered solids: Additional numerical results. *Z. Phys. B* **1983**, *53*, 1–13.10.1007/BF01578242. [CrossRef]

59. Dwivedi, V.; Chua, V. Of bulk and boundaries: Generalized transfer matrices for tight-binding models. *Phys. Rev. B* **2016**, *93*, 134304.10.1103/PhysRevB.93.134304. [CrossRef]

60. Lambropoulos, K.; Simserides, C. Spectral and transmission properties of periodic 1D tight-binding lattices with a generic unit cell: An analysis within the transfer matrix approach. *J. Phys. Commun.* **2018**, *2*, 035013.10.1088/2399-6528/aab065. [CrossRef]

61. Datta, S. *Electronic Transport in Mesoscopic Systems*; Cambridge University Press: Cambridge, UK, 1995.10.1017/CBO9780511805776. [CrossRef]

62. Hawke, L.G.D.; Kalosakas, G.; Simserides, C. Electronic parameters for charge transfer along DNA. *Eur. Phys. J. E* **2010**, *32*, 291–305.10.1140/epje/i2010-10650-y. [CrossRef]

63. Capobianco, A.; Landi, A.; Peluso, A. Modeling DNA oxidation in water. *Phys. Chem. Chem. Phys.* **2017**, *19*, 13571–13578.10.1039/C7CP02029E. [CrossRef]

64. Sugiyama, H.; Saito, I. Theoretical Studies of GG-Specific Photocleavage of DNA via Electron Transfer: Significant Lowering of Ionization Potential and 5′-Localization of HOMO of Stacked GG Bases in B-Form DNA. *J. Am. Chem. Soc.* **1996**, *118*, 7063–7068.10.1021/ja9609821. [CrossRef]

65. Hutter, M.; Clark, T. On the Enhanced Stability of the Guanine-Cytosine Base-Pair Radical Cation. *J. Am. Chem. Soc.* **1996**, *118*, 7574–7577.10.1021/ja953370+. [CrossRef]

66. Zhang, H.; Li, X.Q.; Han, P.; Yu, X.Y.; Yan, Y. A partially incoherent rate theory of long-range charge transfer in deoxyribose nucleic acid. *J. Chem. Phys.* **2002**, *117*, 4578–4584.10.1063/1.1497162. [CrossRef]

67. Li, X.; Cai, Z.; Sevilla, M.D. Investigation of Proton Transfer within DNA Base Pair Anion and Cation Radicals by Density Functional Theory (DFT). *J. Phys. Chem. B* **2001**, *105*, 10115–10123.10.1021/jp012364z. [CrossRef]

68. Li, X.; Cai, Z.; Sevilla, M.D. Energetics of the Radical Ions of the AT and AU Base Pairs: A Density Functional Theory (DFT) Study. *J. Phys. Chem. A* **2002**, *106*, 9345–9351.10.1021/jp021322n. [CrossRef]

69. Shukla, M.K.; Leszczynski, J. A Theoretical Study of Excited State Properties of Adenine-Thymine and Guanine-Cytosine Base Pairs. *J. Phys. Chem. A* **2002**, *106*, 4709–4717.10.1021/jp014516w. [CrossRef]

70. Mantela, M.; Morphis, A.; Tassi, M.; Simserides, C. Lowest ionisation and excitation energies of biologically important heterocyclic planar molecules. *Mol. Phys.* **2016**, *114*, 709–718.10.1080/00268976.2015.1113313. [CrossRef]

71. Voityuk, A.A.; Jortner, J.; Bixon, M.; Rösch, N. Electronic coupling between Watson–Crick pairs for hole transfer and transport in desoxyribonucleic acid. *J. Chem. Phys.* **2001**, *114*, 5614–5620.10.1063/1.1352035. [CrossRef]

72. Ivanova, A.; Shushkov, P.; Rösch, N. Systematic Study of the Influence of Base-Step Parameters on the Electronic Coupling between Base-Pair Dimers: Comparison of A-DNA and B-DNA Forms. *J. Phys. Chem. A* **2008**, *112*, 7106–7114.10.1021/jp8031513. [CrossRef]

73. Migliore, A.; Corni, S.; Varsano, D.; Klein, M.L.; Di Felice, R. First, Principles Effective Electronic Couplings for Hole Transfer in Natural and Size-Expanded DNA. *J. Phys. Chem. B* **2009**, *113*, 9402–9415.10.1021/jp904295q. [CrossRef]

74. Shechtman, D.; Blech, I.; Gratias, D.; Cahn, J.W. Metallic Phase with Long-Range Orientational Order and No Translational Symmetry. *Phys. Rev. Lett.* **1984**, *53*, 1951–1953.10.1103/PhysRevLett.53.1951. [CrossRef]

75. International Union of Crystallography. Report of the Executive Committee for 1991. *Acta Crystallogr. A* **1992**, *48*, 922–946.10.1107/s0108767392008328. [CrossRef]

76. Maciá, E. On the Nature of Electronic Wave Functions in One-Dimensional Self-Similar and Quasiperiodic Systems. *ISRN Condens. Matter Phys.* **2014**, *2014*, 165943.10.1155/2014/165943. [CrossRef]

77. Kosuri, S.; Church, G.M. Large-scale de novo DNA synthesis: technologies and applications. *Nat. Methods* **2014**, *11*, 499–507.10.1038/nmeth.2918. [CrossRef]

78. Beaucage, S.; Caruthers, M. Deoxynucleoside phosphoramidites—A new class of key intermediates for deoxypolynucleotide synthesis. *Tetrahedron Lett.* **1981**, *22*, 1859–1862.10.1016/S0040-4039(01)90461-7. [CrossRef]

79. Palluk, S.; Arlow, D.H.; de Rond, T.; Barthel, S.; Kang, J.S.; Bector, R.; Baghdassarian, H.M.; Truong, A.N.; Kim, P.W.; Singh, A.K.; et al. De novo DNA synthesis using polymerase-nucleotide conjugates. *Nat. Biotechnol.* **2018**, *36*, 645–650.10.1038/nbt.4173. [CrossRef]
80. Perron, O. Zur Theorie der Matrices. *Math. Ann.* **1907**, *64*, 248–263, doi:10.1007/bf01449896. (In German) [CrossRef]
81. Frobenius, G. *Über Matrizen aus nicht Negativen Elementen*; Sitzungsberichte Königlich Preussischen Akademie Wissenschaften Berlin: Berlin, Germany, 1912; pp. 456–477. (In German)
82. Baake, M.; Grimm, U. *Aperiodic Order: Volume 1, A Mathematical Invitation*; Cambridge University Press: Cambridge, UK, 2013.
83. Lambropoulos, K.; Simserides, C. Periodic, quasiperiodic, fractal, Kolakoski, and random binary polymers: Energy structure and carrier transport. *Phys. Rev. E* **2019**, *99*, 032415.10.1103/PhysRevE.99.032415. [CrossRef]
84. Berthé, V.; Siegel, A. Basic notions on substitutions. In *Substitutions in Dynamics, Arithmetics and Combinatorics*; Fogg, N.P., Berthé, V., Ferenczi, S., Mauduit, C., Siegel, A., Eds.; Springer: Berlin/Heidelberg, Germany, 2002; pp. 1–32, doi:10.1007/3-540-45714-3_1.
85. Canterini, V.; Siegel, A. Geometric Representation of Substitutions of Pisot Type. *Trans. Am. Math. Soc.* **2001**, *353*, 5121–5144. [CrossRef]
86. Bombieri, E.; Taylor, J.E. Which distributions of matter diffract? An initial investigation. *J. Phys. Colloq.* **1986**, *47*, 19–28.10.1051/jphyscol:1986303. [CrossRef]
87. Bombieri, E.; Taylor, J.E. Quasicrystals, Tilings, and Algebraic Number Theory: Some Preliminary Connections. In *The Legacy of Sonya Kovalevskaya*; Keen, L., Ed.; Contemporary Mathematics, American Mathematical Society: Providence, RI, USA, 1987; Volume 64, pp. 241–264.
88. Solomyak, B. Dynamics of Self-Similar Tilings. *Ergod. Theory Dyn. Syst.* **1997**, *17*, 695–738.10.1017/S0143385797084988. [CrossRef]
89. Luck, J.M.; Godrèche, C.; Janner, A.; Janssen, T. The nature of the atomic surfaces of quasiperiodic self-similar structures. *J. Phys. A Math. Gen.* **1993**, *26*, 1951–1999.10.1088/0305-4470/26/8/020. [CrossRef]
90. Kolář, M. New class of one-dimensional quasicrystals. *Phys. Rev. B* **1993**, *47*, 5489–5492.10.1103/PhysRevB.47.5489. [CrossRef]
91. Kolář, M.; Iochum, B.; Raymond, L. Structure factor of 1D systems (superlattices) based on two-letter substitution rules. I. delta (Bragg) peaks. *J. Phys. A Math. Gen.* **1993**, *26*, 7343–7366.10.1088/0305-4470/26/24/011. [CrossRef]
92. Sigler, L. *Fibonacci's Liber Abaci: A Translation into Modern English of Leonardo Pisano's Book of Calculation*; Springer: New York, NY, USA, 2003.
93. Birch, J.; Severin, M.; Wahlström, U.; Yamamoto, Y.; Radnoczi, G.; Riklund, R.; Sundgren, J.E.; Wallenberg, L.R. Structural characterization of precious-mean quasiperiodic Mo/V single-crystal superlattices grown by dual-target magnetron sputtering. *Phys. Rev. B* **1990**, *41*, 10398–10407.10.1103/PhysRevB.41.10398. [CrossRef]
94. Fu, X.; Liu, Y.; Zhou, P.; Sritrakool, W. Perfect self-similarity of energy spectra and gap-labeling properties in one-dimensional Fibonacci-class quasilattices. *Phys. Rev. B* **1997**, *55*, 2882–2889.10.1103/PhysRevB.55.2882. [CrossRef]
95. Maciá, E. Spectral Classification of One-Dimensional Binary Aperiodic Crystals: An Algebraic Approach. *Ann. Phys.* **2017**, *529*, 1700079, doi:10.1002/andp.201700079. [CrossRef]
96. Baake, M.; Grimm, W.; Mañibo, N. Spectral analysis of a family of binary inflation rules. *Lett. Math. Phys.* **2018**, *108*, 1783–1805.10.1007/s11005-018-1045-4. [CrossRef]
97. Prouhet, E. Mémoire sur quelques relations entre les puissances des nombres. *CR Acad. Sci. Paris* **1851**, *33*, 225. (In French)
98. Nagell, T.; Selberg, A.; Selberg, S.; Thalberg, K. (Eds.) *Selected Mathematical Papers of Axel Thue*; Universitetsforlaget: Oslo, Norway, 1977.
99. Morse, H.M. Recurrent Geodesics on a Surface of Negative Curvature. *Trans. Am. Math. Soc.* **1906**, *22*, 84–100.10.2307/1988844. [CrossRef]
100. Rudin, W. Some theorems on Fourier coefficients. *Proc. Am. Math. Soc.* **1959**, *10*, 855–895.10.1090/s0002-9939-1959-0116184-5. [CrossRef]
101. Shapiro, H.S. Extremal Problems for Polynomials and Power Series. Master's Thesis, Massachusetts Institute of Technology, Cambridge, MA, USA, 1951.
102. Cantor, G. Über unendliche, lineare Punktmannigfaltigkeiten. *Math. Ann.* **1883**, *21*, 545–586. (In German) [CrossRef]

103. Vaidya, A.M.; Simon, H.; Garrett, P.; Shapiro, H.S.; Kolakoski, W.; Dapkus, F.; Gross, F.; Cohen, M.J.; Comtet, L.; Feller, E.H. Advanced Problems: 5300–5309. *Am. Math. Mon.* **1965**, *72*, 673–675, doi:10.2307/2313883. [CrossRef]
104. Oldenburger, R. Exponent trajectories in symbolic dynamics. *Trans. Am. Math. Soc.* **1939**, *46*, 453–466.10.2307/1989933. [CrossRef]
105. Sing, B. Kolakoski sequences—An example of aperiodic order. *J. Non-Cryst. Solids* **2004**, *334*, 100–104.10.1016/j.jnoncrysol.2003.11.021. [CrossRef]
106. Sing, B. Kolakoski-(2m,2n) are limit-periodic model sets. *J. Math. Phys.* **2003**, *44*, 899–912.10.1063/1.1521239. [CrossRef]
107. Bovier, A.; Ghez, J.M. Remarks on the spectral properties of tight-binding and Kronig-Penney models with substitution sequences. *J. Phys. A Math. Gen.* **1995**, *28*, 2313–2324.10.1088/0305-4470/28/8/022. [CrossRef]
108. Bellisard, J.; Bovier, A.; Ghez, J.M. Gap Labelling Theorems for One Dimensional Discrete Schrödinger operators. *Rev. Math. Phys.* **1992**, *4*, 1–37.10.1142/S0129055X92000029. [CrossRef]
109. Bellisard, J.; Bovier, A.; Ghez, J.M. Discrete Schrödinger Operators with Potentials Generated by Substitutions. In *Differential Equations with Applications to Mathematical Physics*; Ames, W.F., Harrell, E.M., Herod, J.V., Eds.; Elsevier: Amsterdam, The Netherlands, 1993; Volume 192, pp. 13–23, doi:10.1016/S0076-5392(08)62368-1.
110. De Oliveira, B.P.W.; Albuquerque, E.L.; Vasconcelos, M.S. Electronic density of states in sequence dependent DNA molecules. *Surf. Sci.* **2006**, *600*, 3770–3774.10.1016/j.susc.2006.01.081. [CrossRef]
111. De Moura, F.A.B.F.; Lyra, M.L.; Albuquerque, E.L. Electronic transport in poly(CG) and poly(CT) DNA segments with diluted base pairing. *J. Phys. Condens. Matter* **2008**, *20*, 075109.10.1088/0953-8984/20/7/075109. [CrossRef]
112. De Moura, F.A.B.F.; Lyra, M.L.; de Almeida, M.L.; Ourique, G.S.; Fulco, U.L.; Albuquerque, E.L. Methylation effect on the ohmic resistance of a poly-GC DNA-like chain. *Phys. Lett. A* **2016**, *380*, 3559–3563.10.1016/j.physleta.2016.07.069. [CrossRef]
113. Maciá Barber, E. *Aperiodic Structures in Condensed Matter: Fundamentals and Applications*; CRC Press: Boca Raton, FL, USA, 2008.10.1201/9781420068283. [CrossRef]
114. Maciá, E.; Triozon, F.; Roche, S. Contact-dependent effects and tunneling currents in DNA molecules. *Phys. Rev. B* **2005**, *71*, 113106.10.1103/PhysRevB.71.113106. [CrossRef]
115. Mason, J.; Handscomb, D.C. *Chebyshev Polynomials*; Chapman and Hall/CRC: London, UK, 2002.
116. Malakooti, S.; Hedin, E.R.; Kim, Y.D.; Joe, Y.S. Enhancement of charge transport in DNA molecules induced by the next nearest-neighbor effects. *J. Appl. Phys.* **2012**, *112*, 094703.10.1063/1.4764310. [CrossRef]
117. Landauer, R. Spatial Variation of Currents and Fields Due to Localized Scatterers in Metallic Conduction. *IBM J. Res. Dev.* **1957**, *1*, 223–231.10.1147/rd.13.0223. [CrossRef]
118. Buttiker, M. Symmetry of electrical conduction. *IBM J. Res. Dev.* **1988**, *32*, 317–334.10.1147/rd.323.0317. [CrossRef]
119. Roche, S.; Maciá, E. Electronic Transport and Thermopower in Aperiodic DNA Sequences. *Mod. Phys. Lett. B* **2004**, *18*, 847–871.10.1142/S021798490400744X. [CrossRef]
120. Plazas, C.A.; Fonseca-Romero, K.M.; Rey-González, R.R. Insulator-to-Semiconductor-to-Conductor Phase-Like Transition of DNA Chains. *J. Nanosci. Nanotechnol.* **2018**, *18*, 5042–5048.10.1166/jnn.2018.15339. [CrossRef]
121. Kundu, S.; Karmakar, S.N. Conformation dependent electronic transport in a DNA double-helix. *AIP Adv.* **2015**, *5*, 107122, doi:10.1063/1.4934507. [CrossRef]
122. Joe, Y.S.; Lee, S.H.; Hedin, E.R.; Kim, Y.D. Temperature and Magnetic Field Effects on Electron Transport Through DNA Molecules in a Two-Dimensional Four-Channel System. *J. Nanosci. Nanotechnol.* **2013**, *13*, 3889–3896.10.1166/jnn.2013.7206. [CrossRef]
123. De Almeida, M.L.; Oliveira, J.I.N.; Lima Neto, J.X.; Gomes, C.E.M.; Fulco, U.L.; Albuquerque, E.L.; Freire, V.N.; Caetano, E.W.S.; de Moura, F.A.B.F.; Lyra, M.L. Electronic transport in methylated fragments of DNA. *Appl. Phys. Lett.* **2015**, *107*, 203701, doi:10.1063/1.4936099. [CrossRef]
124. Oliveira, J.I.N.; Albuquerque, E.L.; Fulco, U.L.; Mauriz, P.W.; Sarmento, R.G.; Caetano, E.W.S.; Freire, V.N. Conductance of single microRNAs chains related to the autism spectrum disorder. *EPL* **2014**, *107*, 68006, doi:10.1209/0295-5075/107/68006. [CrossRef]

125. Sarmento, R.G.; Silva, R.N.O.; Madeira, M.P.; Frazão, N.F.; Sousa, J.O.; Macedo-Filho, A. Electronic Transport in Single-Stranded DNA Molecule Related to Huntington's Disease. *Braz. J. Phys.* **2018**, *48*, 155–159.10.1007/s13538-018-0554-z. [CrossRef]
126. Walker, F.O. Huntington's disease. *Lancet* **2007**, *369*, 218–228.10.1016/S0140-6736(07)60111-1. [CrossRef]

© 2019 by the authors. Licensee MDPI, Basel, Switzerland. This article is an open access article distributed under the terms and conditions of the Creative Commons Attribution (CC BY) license (http://creativecommons.org/licenses/by/4.0/).

Review

Aperiodic-Order-Induced Multimode Effects and Their Applications in Optoelectronic Devices

Hao Jing, Jie He, Ru-Wen Peng * and Mu Wang

National Laboratory of Solid State Microstructures, School of Physics, and Collaborative Innovation Center of Advanced Microstructures, Nanjing University, Nanjing 210093, China
* Correspondence: rwpeng@nju.edu.cn

Received: 3 August 2019; Accepted: 20 August 2019; Published: 4 September 2019

Abstract: Unlike periodic and random structures, many aperiodic structures exhibit unique hierarchical natures. Aperiodic photonic micro/nanostructures usually support optical multimodes due to either the rich variety of unit cells or their hierarchical structure. Mainly based on our recent studies on this topic, here we review some developments of aperiodic-order-induced multimode effects and their applications in optoelectronic devices. It is shown that self-similarity or mirror symmetry in aperiodic micro/nanostructures can lead to optical or plasmonic multimodes in a series of one-dimensional/two-dimensional (1D/2D) photonic or plasmonic systems. These multimode effects have been employed to achieve optical filters for the wavelength division multiplex, open cavities for light–matter strong coupling, multiband waveguides for trapping "rainbow", high-efficiency plasmonic solar cells, and transmission-enhanced plasmonic arrays, etc. We expect that these investigations will be beneficial to the development of integrated photonic and plasmonic devices for optical communication, energy harvesting, nanoantennas, and photonic chips.

Keywords: quasiperiodic order; self-similarity; quasiperiodic photonic micro/nanostructures; fractal-like photonic micro/nanostructures; quasiperiodic or fractal-like plasmonic structures

1. Introduction

Motivated by the discovery of quasicrystal [1], much research has been conducted on aperiodic systems in recent years [2–11]. Aperiodic structures broaden the regime of ordered systems beyond periodic structures, and thereby play a significant role in a wide range of science and engineering disciplines. Unlike periodic and random structures, many aperiodic structures exhibit unique hierarchical natures. For these aperiodic structures, on one hand, the lack of periodicity may create fascinating features on some occasions, such as extraordinary optical transmission and enhanced transmission resonances, etc. On the other hand, aperiodic order can be artificially imposed during sample fabrication and can be precisely controlled. These properties have opened a new avenue for the design of novel devices based on aperiodic structures. Among them, optoelectronic devices based on the multimode effects, which can be induced by aperiodic order, have attracted much attention because of their potential in optical communication [12,13], energy harvesting [14], nanoantennas [15], and so on.

In this review, we mainly summarize the research work in our group concerning the aperiodic-order-induced multimode effects, which have been demonstrated by both theoretical and experimental observations. Moreover, several optoelectronic devices have been designed on the basis of the multimode effects in either one-dimensional (1D) or two-dimensional (2D) aperiodic structures.

2. Aperiodic Structures

In order to design optoelectronic devices based on aperiodic structures, proper aperiodic lattices should be chosen. These are usually obtained by following the substitution rules. Several typical aperiodic structures are introduced in this section, ranging from one dimension to two dimensions.

2.1. One-Dimensional Aperiodic Structures

A representative example of 1D aperiodic structures is the Fibonacci structure, which follows Fibonacci sequence as shown in Figure 1a [16]. This sequence can be constructed by applying the substitution rule A→AB and B→A repeatedly. By intentionally varying the growth sequence and the number of building blocks, a standard two-component Fibonacci structure can be generalized to a k-component Fibonacci structure [17]. The feature of this k-component Fibonacci structure is related to k, which can show periodic ($k = 1$), quasiperiodic ($k \leq 5$), or only aperiodic ordering ($k > 5$).

The Thue–Morse sequence is another well-known 1D aperiodic sequence. It can be structured by repeating two building blocks (A and B) applying the substitution rules A→AB and B→BA (Figure 1b) [18]. The initial few generations S_n of the Thue–Morse sequence have the following forms: S_0 = {A}, S_1 = {AB}, S_2 = {ABBA}, S_3 = {ABBABAAB}, and so on. The Thue–Morse lattice is not quasiperiodic but deterministically aperiodic, which shows the properties of an intermediate between periodic and quasiperiodic lattices [19,20].

On the basis of various substitution rules, many other 1D aperiodic sequences have been proposed. For example, the Rudin–Shapiro sequence can be produced by repeated application of the substitution rule AA→BBAB and BB→BBBA [21], and the period-doubling sequence can be structured by using the substitution rule A→AB and B→AA [22]. In addition, a kind of quasiperiodic superlattice structures called the precious mean sequences were reported by Birch et al., which can be produced by A→A^nB and B→A [23], whereas the metallic mean sequences can be generated by the inflation rule A→AB^n and B→A [24].

2.2. Two-Dimensional Aperiodic Structures

Substitution rules used in 1D quasiperiodic structures can be extended to two dimensions. A simple way to obtain a 2D quasiperiodic lattice is to alternate the iterations of 1D inflation rules along different spatial dimensions. For example, as shown in Figure 1c, a 2D Fibonacci quasi-lattice can be structured by using two complementary 1D Fibonacci inflation maps along the horizontal and vertical directions, respectively (f_A: A→AB, B→A; f_B: A→B, B→BA) [25].

Another way to construct 2D quasiperiodic structures is by employing aperiodic tilings. These tilings are composed of collections of polygons, which could cover a plane without gaps and overlaps with a lack of translational symmetries [26]. Various aperiodic tilings have been proposed before, such as Penrose tiling [27] and square Fibonacci tiling [28], etc. Figure 1d illustrates a Penrose tiling; this tiling is composed of two types of rhombuses. A Penrose construction possesses a long-range quasiperiodic order but lacks translational symmetry. In this case, the notion of repetitiveness mainly shows local isomorphism instead of periodic arrangements [8].

2.3. Fractal Patterns

Fractal patterns exhibit self-similarity, where a structure is repeated over multiple spatial scales. Similar to quasi-crystalline order, certain motifs of the self-similar samples contain the whole structure enfolded within them [29]. A typical fractal design is the Koch snowflake fractal. It can be obtained by repeatedly constructing new triangles based on the middle segments of previous triangles; therefore, fractals can be defined by iteration. A triadic Koch snowflake fractal with an iteration of 3 is shown in Figure 1e [30]. Figure 1f shows another typical fractal pattern, i.e., the Sierpinski carpet pattern. It consists of hierarchically-arranged iteratively-shrinking squares, showing different sizes at different scales [31].

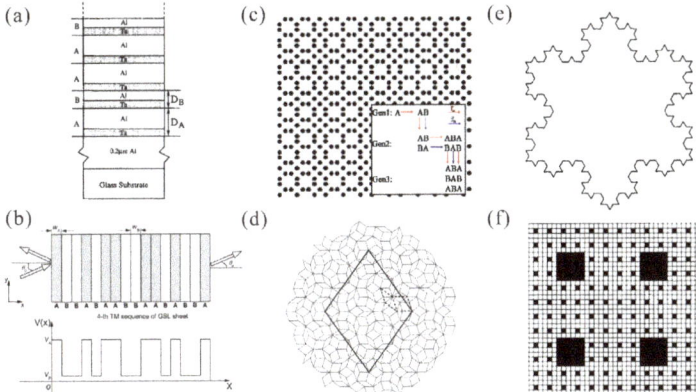

Figure 1. Schematic description of the typical aperiodic structures. (**a**) Fibonacci structure [16]. (**b**) Thue–Morse structure [18]. (**c**) 2D Fibonacci structure [25]. (Inset: Inflation rules of the first two generations of a 2D Fibonacci sequence). (**d**) Regular Penrose tiling [27]. (**e**) Koch snowflake fractal [30]. (**f**) Sierpinski carpet pattern [31].

3. Aperiodic-Order-Induced Multimode Effects in Photonic Micro/Nanostructures

Photons in periodic dielectric structures such as photonic crystals can be considered as a counterpart to electrons in solids. As analog of electronic band structures, photonic band structures possess bandgaps in which photons are prohibited from propagating. In view of the fact that numerous photonic devices are required to work at specific wavelengths or photonic modes, the introduction of photonic band structures with multiple modes is desirable for designing these devices. Studies on the multimode effects have been extended to various aperiodic structures including quasiperiodic structures, symmetric self-similar structures, and others. The multimode effects in aperiodic structures offer a new platform for using photons with various frequencies at the same time.

3.1. Multimode Effects Induced by Self-Similarity

Multiple fundamental photonic band gaps (PBGs) can exist in some aperiodic dielectric multilayers. A typical example is in the Thue–Morse structure (Figure 2a) [32], in which the self-similarity of the structure imparts a trifurcation feature on the resonant transmissions around the central frequency. In the Thue–Morse multilayer, the amount of the completely transparent states can be counted [33]. We define R_n as the amount of the resonant transmission mode around the central frequency (where n represents the number of the generation). Then we obtain:

$$R_{n+2} = R_n + R_{n+1} + R_n \quad (n \geq 3) \tag{1}$$

considering the initial conditions $R_3 = 1$ and $R_4 = 3$. Finally, we have:

$$R_n = 2R_{n-1} - 1 \pm 1 = 1 + \frac{2^{n-1} \pm 1}{3}, \quad where \begin{cases} + \text{ for even } n \\ - \text{ for odd } n \end{cases}, \tag{2}$$

which is the number of PBGs around the central frequency ω_0. According to Equation (2), the inner feature of the Thue–Morse structure determines the mode amount of resonant transmissions. That is to say, special positional correlation between two blocks in the Thue–Morse structure causes resonant transmission. Multiple PBGs can coexist at the same frequency range in these structures, which is intuitively shown in the photonic band structures in Figure 2b. Moreover, the number of PBGs can be increased by tuning the refractive index contrast. This theoretical analysis was verified by measuring Thue–Morse SiO_2/TiO_2 multilayers in the range of visible and near-infrared frequencies [34].

Similar to the Thue–Morse structures, photonic quasicrystals can also support multiple modes because of their self-similarity properties [35,36]. For example, a photonic quasicrystal with eighth-generation Fibonacci series made by two blocks of Ta_2O_5 and SiO_2 was constructed (Figure 2c). From the transmission spectra calculated by the transfer matrix method, three photonic modes could be observed in the dispersion map shown in Figure 2d, as demonstrated experimentally [37].

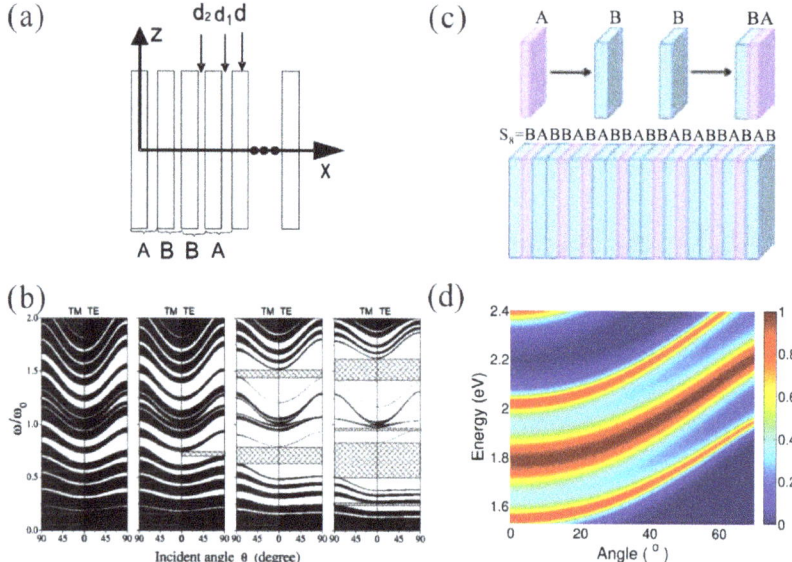

Figure 2. Aperiodic structures and their optical band diagrams: (**a,b**) Thue–Morse multilayer structure [32,34], and (**c,d**) Fibonacci multilayer structure [37].

3.2. Multimode Effects in Symmetric Aperiodic Structure

As mentioned previously, multiple modes appearing in photonic quasiperiodic structures have been demonstrated in both calculations and experiments [17,37,38]. However, low transmission coefficients in these work limit their potential applications. A feasible way to realize multiple perfect transmissions is to introduce internal symmetry into a 1D aperiodic dielectric multilayer structure [39–41]. For example, a photonic crystal with two types of layers can be arranged in a binary Fibonacci-class (FC(n)) sequence. Then, binary symmetric Fibonacci-class (SFC(n)) can be constructed as shown in Figure 3a. SiO_2 and TiO_2 were chosen as two elementary layers with the thickness of a quarter wavelength ($\lambda_0/4$), and the transmission coefficient for the two different systems can be calculated by the transfer matrix method (Figure 3b,c), which shows that the transmission coefficient of the symmetric Fibonacci structure behaves rather differently from that of the Fibonacci structure [39]. As shown in Figure 3, the localization property of optical waves can be influenced by the symmetric internal structure in a quasiperiodic system, which is demonstrated by the sharp transmission peaks with transmission coefficients near unity. That is to say, with the help of symmetric internal structures in the quasiperiodic system, a perfect transmission of the optical wave can replace the initially poor transmission. This improvement is benefited from the positional correlations in the system. Moreover, the resonant transmission can be varied to a certain frequency by tuning the aperiodic structures (Figure 3d). For example, as shown in Figure 3e, the transmission coefficients vary in different symmetric multilayers with defects (SMD) [41].

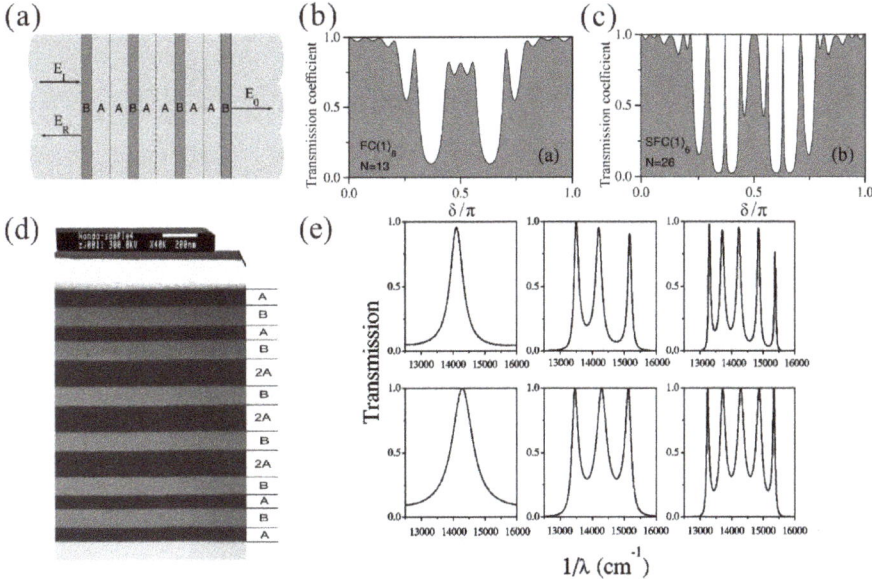

Figure 3. Multimodes in symmetric aperiodic structures. (**a**) Schematic of the symmetric Fibonacci-class (SFC(1)) multilayer structure. (**b**) The transmission coefficient of Fibonacci-class (FC(1)) (13 layers) and (**c**) SFC(1) (26 layers) systems. (**d**) Symmetrical fifth-generation Fibonacci TiO_2/SiO_2 multilayer film. (**e**) The measured (upper row) and calculated (lower row) transmission coefficient T as a function of the wave number for the symmetric TiO_2/SiO_2 mutilayers with defects in the central gap with different layers (SMD V_2, SMD V_3, and SMD V_4 from left to right). (Adapted from ref. [39] (a–c) and ref. [41] (d,e)).

4. Optoelectronic Devices Based on One-Dimensional Aperiodic Structures

The application diversity of modern optoelectronic and photonic devices requires novel functionalities and the tunability of band structures enabled by a unique alignment of materials. By introducing 1D aperiodic order into multilayer structures, some optical modes can be generated at the desired frequencies, which can be applied in constructing functional components such as optical filters, multiband waveguides, and so on.

4.1. Optical Filters for the Wavelength Division Multiplexing (WDM) Systems

The propagation of photons with a certain range of energies can be suppressed by PBGs in photonic crystals. Tunable structural parameters are more plentiful in quasiperiodic designs than those in periodic structures, which can be used to control the propagation of light waves with high transmittivity at desired frequencies. Moreover, by combining with mirror symmetry of the structures, resonant transmission will definitely occur, which makes it possible to fabricate multiwavelength narrow band optical filters. For example, an optical filter could be fabricated by following k-component Fibonacci structures. According to the calculated results, the optical transmission coefficient shows a plentiful structure, which depends on the different incommensurate interval sequences k, the layer number N, and the frequency of the light (Figure 4). The transmission coefficient can be tuned by changing the layer number and the number of k; this property makes it useful in the design of high-performance optical filters [40–42].

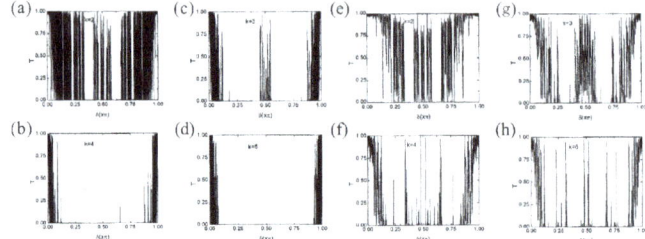

Figure 4. Transmission coefficient T as a function of the phase with the different incommensurate intervals k in k-component Fibonacci structures. The number of layers N are as follows: (**a**) $N = 28,657$; (**b**) $N = 27,201$; (**c**) $N = 31,422$; (**d**) $N = 29,244$; (**e**) $N = 233$; (**f**) $N = 277$; (**g**) $N = 250$; and (**h**) $N = 245$, respectively. (Adapted from ref. [42]).

4.2. Open Cavities for the Light–Matter Strong Coupling

Recently intensive studies have been carried out on light–matter interactions, especially their strong coupling. Apart from the studies on interactions between a single excitonic mode with an individual photonic mode, there has been some work on multimode coupling where the excitonic mode couples with multiple photonic modes. Photonic quasicrystals possess multiple optical modes and thus present a platform for showing multimode light–matter interaction. In order to demonstrate it experimentally, a Fibonacci sequence composed of SiO_2/Ta_2O_5 multilayers was chosen and J-aggregates on the top surface of structure offered excitons (Figure 5a) [37]. Figure 5b shows the measured transmission spectrum, where three peaks of different optical modes are recognized. The Rabi splitting and newly generated hybrid polariton bands can be verified from the dispersion map of the hybrid system, clearly showing successive coupling between the modes H, C, and L and the excitons (Figure 5c,d). By varying the substitution rule of the photonic quasicrystal, the open-cavity system can be optimized to provide the various photonic modes in need. By introducing this design, multimode photon–exciton strong couplings can be realized, which may inspire some potential applications, such as optical spectroscopy and multimode sensors.

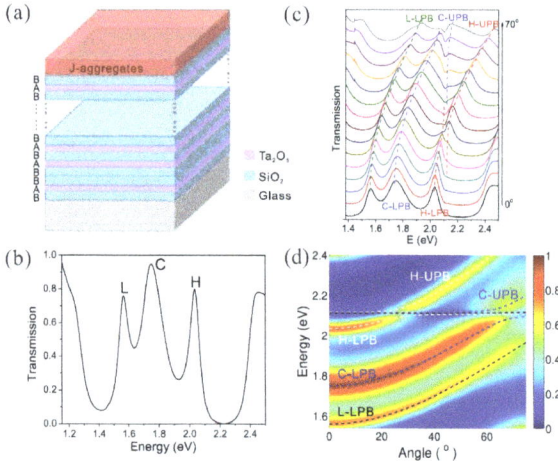

Figure 5. Multimode photon-exciton coupling. (**a**) Schematic of a Fibonacci photonic quasicrystal with J-aggregates on the top surface. (**b**) Experimentally measured transmission spectra of the photonic quasicrystal; the modes labeled C, L, and H correspond to three peaks. (**c**) Transmission spectra of the sample under various incident angles. Polariton bands were traced by dashed lines. (**d**) Dispersion map of the sample. Calculated dashed lines fit the polariton bands and Rabi splitting. (Adapted from ref. [37]).

4.3. Multiband Waveguides for Trapping "Rainbow"

In telecommunications and optoelectronics, optical waveguides play a significant role because of their abilities to confine and guide the light waves. However, conventional hollow-core designs have disadvantages such as narrow transmission bands and detrimental dispersive resonances. Introducing a self-similar dielectric waveguide (SDW) is a useful approach to achieve multiband transmission and overcome baneful dispersive resonance, and even to guide the light waves with spatial separation [43]. As shown in Figure 6a, the SDW is designed as a hollow core surrounded by a coaxial Thue–Morse multilayer. In the photonic band structure, multiple transmission bands appear because of the intrinsic self-similar furcation of the structure. In this case, the propagated light with different resonant frequencies are separated in various cladding layers as shown in Figure 6b. Therefore, different modes are separated spatially, forming a "rainbow" trapped in the SDW (Figure 6c). Moreover, both the transmission modes and the photonic bands can be modulated by altering the temperature in an SDW infiltrated by liquid crystal [44]. These designs can be applied to fabricate compact photonic devices, such as integrated spectrographs, color-sorters, and temperature-sensitive optical circuit switches.

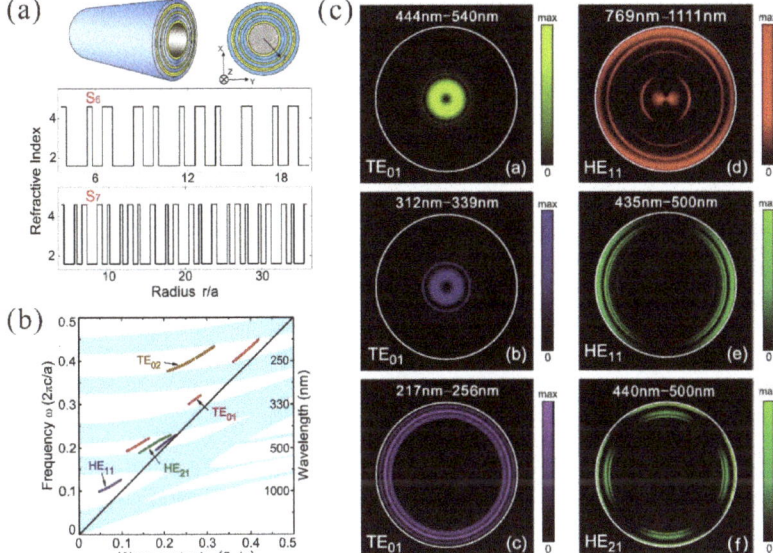

Figure 6. Multiband waveguide. (**a**) Structure of a self-similar dielectric waveguide (SDW), where a coaxial Thue–Morse multilayer consisting of two building blocks was employed to cover a hollow core. The lower figure manifests refractive-index distributions in the SDWs. (**b**) Photonic bands and transmission modes in the SDW. (**c**) The electric-field time-average energy density distribution in the SDW for different modes. (Adapted from ref. [43]).

4.4. Solar Cells with Multi-Intermediate Band Structures

Numerous designs have been developed to improve the performance of solar cells. The enhancement of efficiency may originate from additional photon-induced transitions between the designed intermediate levels, as shown in Figure 7a [45]. Therefore, additional photons whose energies are lower than the original band gap in the solar cell can be absorbed, due to the transitions between bands in the multiband structure. In this way, various intermediate band structures can yield different efficiency limits for solar cells as shown in Figure 7b. It is shown that aperiodic semiconductor superlattices can produce these intermediate energy bands. For example, the continuous minibands in the $In_{0.49}Ga_{0.51}P/GaAs$ superlattices can be split by introducing aperiodic order, such as that of the

Thue–Morse sequence, the Fibonacci sequence, or even the random case (Figure 7c). This approach by introducing a multi-intermediate band structure may produce low-dimensional high-performance photovoltaic devices based on electronic band gap engineering, and can also be used in other ranges such as optoelectronics.

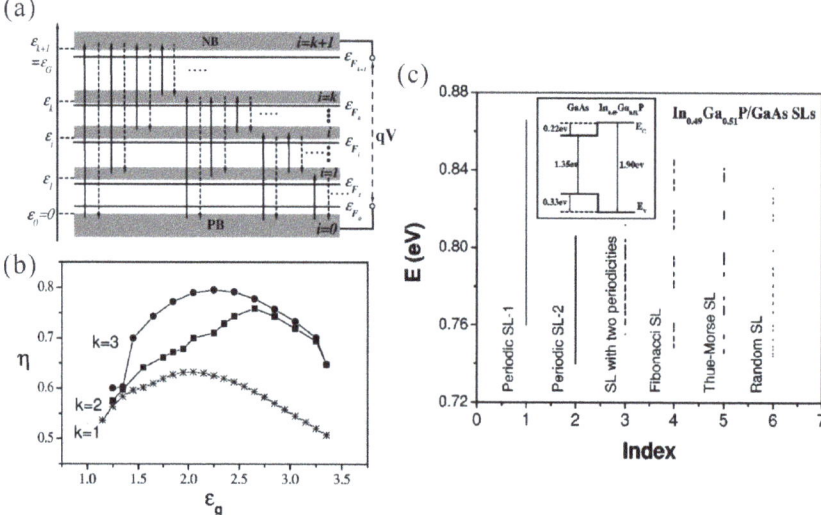

Figure 7. Enhancement of solar cells caused by photon-induced transitions in multi-intermediate band structures. (a) Various radiation transitions between intermediate multiband structures in the designed solar cell. (b) Limiting efficiency η for three model solar cells with diverse intermediate multiband structures (k = 1,2,3). (c) Electronic miniband structures of several periodic and aperiodic $In_{0.49}Ga_{0.51}P$/GaAs superlattices. The inset shows the band-edge diagram of the $In_{0.49}Ga_{0.51}P$/GaAs interface (at room temperature). (Adapted from ref. [45]).

5. Optoelectronic Devices Based on Two-Dimensional Aperiodic Structures

Combining typical 2D aperiodic structures (such as Penrose tiling or fractal patterns) with optoelectronic devices is a feasible way to achieve resonant transmission or absorption enhancement, which can improve the optical response of devices and pave the way toward the integration of devices on a chip.

5.1. Aperiodic Plasmonic Aperture Arrays with Extraordinary Optical Transmission

As we know, much attention has been paid to the resonant transmission of light through subwavelength apertures for its potential applications in photolithography, displays, and near-field microscopy. The phenomena originate from the effect in which surface plasmon polaritons (SPPs) mediate light transmission through the periodic structure. Actually, resonantly enhanced transmission can be achieved not only in periodic structures but also through 2D quasiperiodic aperture arrays (Figure 8a) [46]. The broad transmission of a single aperture can interact with the discrete resonances caused by diffraction from the array, thus the spectral peaks described by Fano interference at terahertz frequencies occur, as shown in Figure 8b.

In addition to quasiperiodic structures, geometric self-similarity can also be employed to make multiple resonant transmission [47,48]. A Sierpinski carpet fractal-featured metallic thin film was fabricated as shown in Figure 8c. The existence of extraordinarily high transmission at specific wavelengths in infrared frequencies was verified by the transmission spectra. This high transmission was determined by the hierarchy of apertures of various sizes. Therefore, this unique structure may play an important role in the miniaturization and integration of plasmonic circuits.

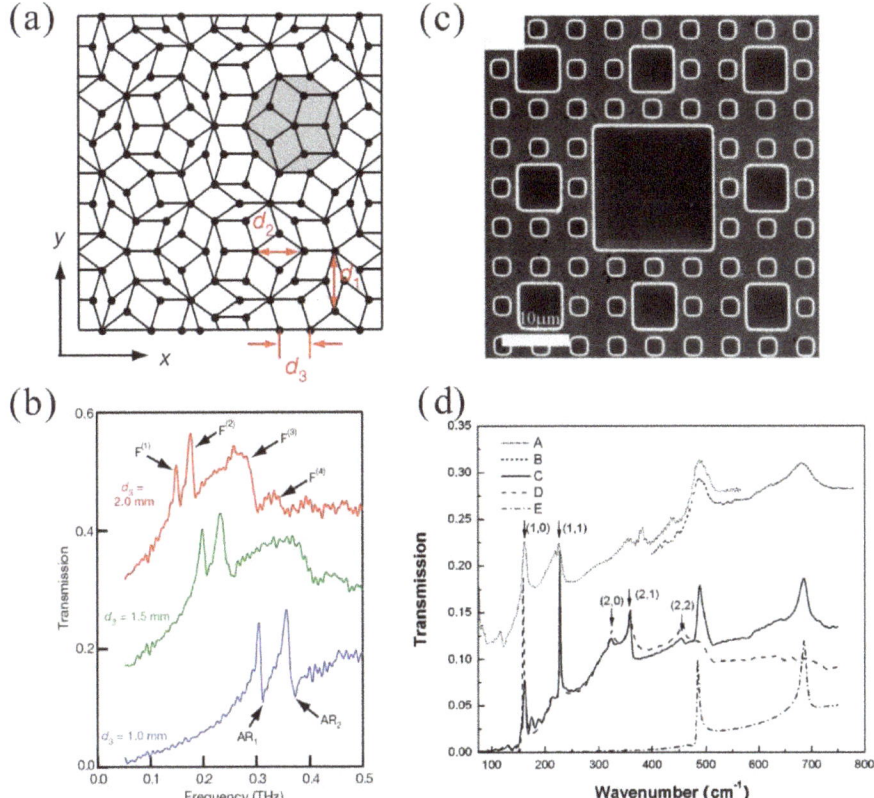

Figure 8. Transmission through aperiodic aperture arrays. (**a**) A Penrose quasicrystal constructed of thin and thick rhomb tiles. (**b**) Transmission of the Penrose-type quasicrystal perforated films with different side lengths. (**c**) Scanning electron micrograph of an aluminum film perforated with a Sierpinski carpet fractal-featured aperture array. (**d**) Transmission of the metallic Sierpinski carpet structure. (a,b) are obtained experimentally; (c–e) are simulated results. (Adapted from ref. [46] (a,b) and ref. [47] (c,d)).

5.2. Solar Cell with a Plasmonic Fractal

The keys to fabricating high-performance solar cells are to extend the absorption of sunlight irradiation to broader bandwidths and increase the power conversion efficiency. A feasible way to enhance broadband absorption is to introduce silver nano cuboids with a fractal-like pattern atop a silicon solar cell, as shown in Figure 9 [49]. The incident light with different wavelengths could couple into various cavity modes and surface plasmon modes in the structure. In this system, the cavity modes originate from Fabry–Perot resonances at the longitudinal and transverse cavities, while the surface plasmon modes exist at the silicon–silver interface. Benefitting from the various feature sizes in the fractal structure, low-index and high-index surface plasmon modes are excited simultaneously. Eventually, broadband absorption can be achieved in this solar cell. By tuning the geometry of the fractal and applying an additional SiO_2 antireflection layer, the quantum efficiency of the solar cell could be improved further. Therefore, these kinds of plasmonic fractal structures can be applied to design miniaturized compact photovoltaic devices with high performance.

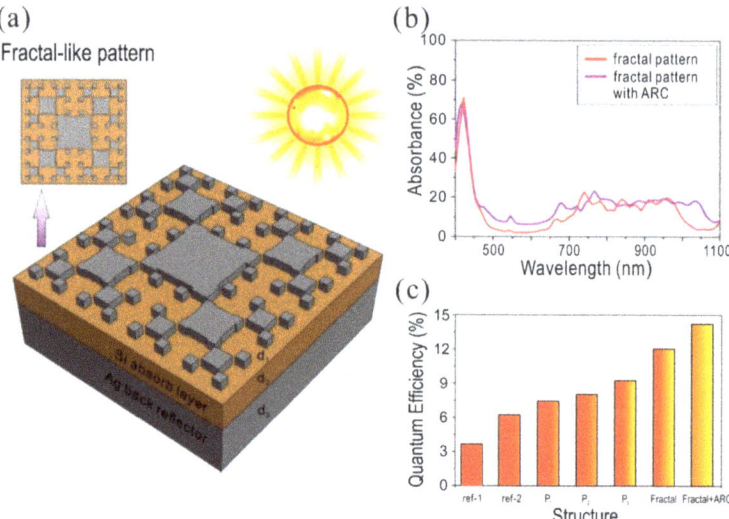

Figure 9. Solar cell with a plasmonic fractal to achieve broadband absorption. (**a**) Schematic of the solar cell with a plasmonic fractal. Including a Ag fractal-like pattern, a Si absorbent layer, and a silver back reflector. (**b**) Absorbance spectra of the solar cell with or without the antireflection coating (ARC). (**c**) Quantum efficiencies of the 50 nm thick silicon solar cells: ref-1 (bare Si film); ref-2 (Si film with Ag back reflector); solar cells with base-periodicity patterns (P1 = 100 nm, P2 = 200 nm, P3 = 400 nm); solar cells with a plasmonic fractal and additional dielectric ARC. (Adapted from ref. [49]).

6. Conclusions

Mainly based on the recent work in our group, we have briefly reviewed the aperiodic-order-induced multimode effects in photonic and plasmonic micro/nanostructures and their applications in optoelectronic devices. We present the multimode effects in a series of 1D/2D photonic or plasmonic aperiodic structures. These multimode effects have been employed to achieve optical filters for the WDM systems, open cavities for light–matter strong coupling, multiband waveguides for trapping "rainbow", high-efficiency plasmonic solar cells, transmission-enhanced plasmonic arrays, and other devices. The investigations can be applied to the design of integrated photonic and plasmonic devices, which achieve potential applications in areas of optical communication, optical data storage, energy harvesting, nanoantennas, photonic chips, and so on.

Author Contributions: R.W.P., M.W., and H.J. conceived the idea. H.J., J.H., and R.W.P. wrote the paper. R.W.P. and M.W. supervised the research.

Funding: This research was funded by the National Key R&D Program of China (2017YFA0303702) and the National Natural Science Foundation of China (Grants No. 11634005 and No. 11674155).

Conflicts of Interest: The authors declare no conflict of interest.

References

1. Shechtman, D.; Blech, I.; Gratias, D.; Cahn, J.W. Metallic phase with long-range orientational order and no translational symmetry. *Phys. Rev. Lett.* **1984**, *53*, 1951. [CrossRef]
2. Merlin, R.; Bajema, K.; Clarke, R. Quasiperiodic GaAs-AlAs heterostructures. *Phys. Rev. Lett.* **1985**, *55*, 1768–1770. [CrossRef]

3. Todd, J.; Merlin, R.; Clarke, R.; Mohanty, K.M.; Axe, J.D. Synchrotron X-ray study of a Fibonacci superlattice. *Phys. Rev. Lett.* **1986**, *57*, 1157–1160. [CrossRef]
4. Hu, A.; Tien, C.; Li, X.J.; Wang, Y.H.; Feng, D. X-ray diffraction pattern of quasiperiodic (Fibonacci) Nb-Cu superlattices. *Phys. Lett. A* **1986**, *119*, 313–314. [CrossRef]
5. Kohmoto, M.; Sutherland, B.; Iguchi, K. Localization in optics: Quasiperiodic media. *Phys. Rev. Lett.* **1987**, *58*, 2436–2438. [CrossRef]
6. Dharma-wardana, M.W.C.; MacDonald, A.H.; Lockwood, D.J.; Baribeau, J.M.; Houghton, D.C. Raman scattering in Fibonacci superlattices. *Phys. Rev. Lett.* **1987**, *58*, 1761. [CrossRef]
7. Gellermann, W.; Kohmoto, M.; Sutherland, B.; Taylor, P.C. Localization of light waves in Fibonacci dielectric multilayers. *Phys. Rev. Lett.* **1994**, *72*, 633–636. [CrossRef]
8. Maciá, E. Exploiting aperiodic designs in nanophotonic devices. *Rep. Prog. Phys.* **2012**, *75*, 036502. [CrossRef]
9. Grimm, U. Aperiodic crystals and beyond. *Acta Crystallogr. B* **2015**, *71*, 258–274. [CrossRef]
10. Peng, R.W.; Hu, A.; Jiang, S.S. Study on quasiperiodic Ta/Al multilayer films by x-ray diffraction. *Appl. Phys. Lett.* **1991**, *59*, 2512. [CrossRef]
11. Peng, R.W.; Hu, A.; Jiang, S.S.; Zhang, C.S.; Feng, D. Structural characterization of 3-component Fibonacci Ta/Al multilayer films. *Phys. Rev. B* **1992**, *46*, 7816. [CrossRef] [PubMed]
12. Chakraborty, S.; Marshall, O.P.; Folland, T.G.; Kim, Y.J.; Grigorenko, A.N.; Novoselov, K.S. Gain modulation by graphene plasmons in aperiodic lattice lasers. *Science* **2015**, *351*, 246–248. [CrossRef] [PubMed]
13. Verslegers, L.; Catrysse, P.B.; Yu, Z.; Fan, S. Deep-Subwavelength Focusing and Steering of Light in an Aperiodic MetallicWaveguide Array. *Phys. Rev. Lett.* **2009**, *103*, 033902. [CrossRef] [PubMed]
14. Pala, R.A.; Liu, J.S.; Barnard, E.S.; Askarov, D.; Garnett, E.C.; Fan, S.; Brongersma, M.L. Optimization of non-periodic plasmonic light-trapping layers for thin-film solar cells. *Nat. Commun.* **2013**, *4*, 2095. [CrossRef] [PubMed]
15. Fan, J.A.; Yeo, W.H.; Su, Y.; Hattori, Y.; Lee, W.; Jung, S.Y.; Zhang, Y.; Liu, Z.; Cheng, H.; Falgout, L.; et al. Fractal design concepts for stretchable electronics. *Nat. Commun.* **2014**, *5*, 3266. [CrossRef] [PubMed]
16. Jiang, S.S.; Hu, A.; Peng, R.W.; Feng, D. Quasiperiodic metallic multilayers. *J. Magn. Magn. Mater.* **1993**, *162*, 82–88. [CrossRef]
17. Peng, R.W.; Wang, M.; Hu, A.; Jiang, S.S.; Jin, G.J.; Feng, D. Characterization of the diffraction spectra of one-dimensional *k*-component Fibonacci structures. *Phys. Rev. B* **1995**, *52*, 13310–13316. [CrossRef]
18. Ma, T.; Liang, C.; Wang, L.; Lin, H.Q. Electronic band gaps and transport in aperiodic graphene superlattices of Thue–Morse sequence. *Appl. Phys. Lett.* **2012**, *100*, 252402. [CrossRef]
19. Li, Y.; Peng, R.W.; Jin, G.J.; Wang, M.; Huang, X.Q.; Hu, A.; Jiang, S.S. Persistent currents in one-dimensional aperiodic mesoscopic rings. *Eur. Phys. J. B* **2002**, *25*, 497–503. [CrossRef]
20. Ryu, C.S.; Oh, G.Y.; Lee, M.H. Extended and critical wave functions in a Thue–Morse chain. *Phys. Rev. B* **1992**, *46*, 5162–5168. [CrossRef]
21. Kola, M.; Nori, F. Trace maps of general substitutional sequences. *Phys. Rev. B* **1990**, *42*, 1062–1065. [CrossRef] [PubMed]
22. Bellissard, J.; Bovier, A.; Ghez, J.M. Spectral properties of a tight binding Hamiltonian with Period doubling potential. *Commun. Math. Phys.* **1991**, *135*, 379–399. [CrossRef]
23. Birch, J.; Severin, M.; Wahlstrom, U.; Yamamoto, Y.; Radnoczi, G.; Riklund, R.; Sundgren, J.; Wallenberg, L.R. Structural characterization of precious-mean quasiperiodic Mo/V single-crystal superlattices grown by dual-target magnetron sputtering. *Phys. Rev. B* **1990**, *41*, 10398–10407. [CrossRef] [PubMed]
24. Dotera, T.; Bekku, S.; Ziherl, P. Bronze-mean hexagonal quasicrystal. *Nat. Mater.* **2017**, *16*, 987–992. [CrossRef] [PubMed]
25. Dallapiccola, R.; Gopinath, A.; Stellacci, F.; Negro, L.D. Quasi-periodic distribution of plasmon modes in two-dimensional Fibonacci arrays of metal nanoparticles. *Opt. Express* **2008**, *16*, 5544–5555. [CrossRef] [PubMed]
26. Pierro, V.; Galdi, V.; Castaldi, G.; Pinto, I.M.; Felsen, L.B. Radiation properties of planar antenna arrays based on certain categories of aperiodic tilings. *IEEE Trans. Antennas Propag.* **2005**, *53*, 635–644. [CrossRef]
27. Wang, K. Structural effects on light wave behavior in quasiperiodic regular and decagonal Penrose-tiling dielectric media: A comparative study. *Phys. Rev. B* **2007**, *76*, 085107. [CrossRef]
28. Lifshitz, R. The square Fibonacci tiling. *J. Alloys Compd.* **2002**, *342*, 186–190. [CrossRef]
29. Maciá, E. The role of aperiodic order in science and technology. *Rep. Prog. Phys.* **2006**, *69*, 397–441. [CrossRef]

30. Uozumi, J.; Kimura, H.; Asakura, T. Fraunhofer Diffraction by Koch Fractals. *J. Mod. Opt.* **1990**, *37*, 1011–1031. [CrossRef]
31. Gefen, Y.; Meir, Y.; Mandelbrot, B.B.; Aharony, A. Geometric implementation of hypercubic lattices with noninteger dimensionality by use of low lacunarity fractal lattices. *Phys. Rev. Lett.* **1983**, *50*, 145–148. [CrossRef]
32. Zhang, X.F.; Peng, R.W.; Kang, S.S.; Cao, L.S.; Zhang, R.L.; Wang, M.; Hu, A. Tunable High-frequency Magnetostatic Waves in Thue–Morse Antiferromagnetic Multilayers. *J. Appl. Phys.* **2006**, *100*, 063911. [CrossRef]
33. Qiu, F.; Peng, R.W.; Huang, X.Q.; Liu, Y.M.; Wang, M.; Hu, A.; Jiang, S.S. Resonant transmission and frequency trifurcation of light waves in Thue–Morse dielectric multilayers. *Europhys. Lett.* **2003**, *63*, 853–859. [CrossRef]
34. Qiu, F.; Peng, R.W.; Huang, X.Q.; Hu, X.F.; Wang, M.; Hu, A.; Jiang, S.S.; Feng, D. Omnidirectional reflection of electromagnetic waves on Thue–Morse dielectric multilayers. *Europhys. Lett.* **2004**, *68*, 658–663. [CrossRef]
35. Huang, X.; Wang, Y.; Gong, C. Numerical investigation of light-wave localization in optical Fibonacci superlattices with symmetric internal structure. *J. Phys. Condens. Matter* **1999**, *11*, 517–520. [CrossRef]
36. Peng, R.W.; Jin, G.J.; Wang, M.; Hu, A.; Jiang, S.S.; Feng, D. Interface optical phonons in *k*-component Fibonacci dielectric multilayers. *Phys. Rev. B* **1999**, *59*, 3599–3605. [CrossRef]
37. Zhang, K.; Xu, Y.; Chen, T.Y.; Jing, H.; Shi, W.B.; Xiong, B.; Peng, R.W.; Wang, M. Multimode photon-exciton coupling in an organic-dye-attached photonic quasicrystal. *Opt. Lett.* **2016**, *41*, 5740–5743. [CrossRef]
38. Hu, A.; Wen, Z.X.; Jiang, S.S.; Tong, W.T.; Peng, R.W.; Feng, D. One-dimensional *k*-component Fibonacci Structures. *Phys. Rev. B* **1993**, *48*, 829–835. [CrossRef]
39. Huang, X.Q.; Jiang, S.S.; Peng, R.W.; Hu, A. Perfect transmission and self-similar optical transmission spectra in symmetric Fibonacci-class multilayers. *Phys. Rev. B* **2001**, *63*, 245104. [CrossRef]
40. Peng, R.W.; Huang, X.Q.; Qiu, F.; Wang, M.; Hu, A.; Jiang, S.S.; Mazzer, M. Symmetry-induced perfect transmission of light waves in quasiperiodic dielectric multilayers. *Appl. Phys. Lett.* **2002**, *80*, 3063–3065. [CrossRef]
41. Peng, R.W.; Liu, Y.M.; Huang, X.Q.; Qiu, F.; Wang, M.; Hu, A.; Jiang, S.S.; Feng, D.; Ouyang, L.Z.; Zou, J. Dimerlike positional correlation and resonant transmission of electromagnetic waves in aperiodic dielectric multilayers. *Phys. Rev. B* **2004**, *69*, 165109. [CrossRef]
42. Peng, R.W.; Wang, M.; Hu, A.; Jiang, S.S.; Jin, G.J.; Feng, D. Photonic localization in one-dimensional *k*-component Fibonacci structures. *Phys. Rev. B* **1998**, *57*, 1544. [CrossRef]
43. Hu, Q.; Zhao, J.Z.; Peng, R.W.; Gao, F.; Zhang, R.L.; Wang, M. "Rainbow" trapped in a self-similar coaxial optical waveguide. *Appl. Phys. Lett.* **2010**, *96*, 161101. [CrossRef]
44. Hu, Q.; Xu, D.H.; Peng, R.W.; Zhou, Y.; Yang, Q.L.; Wang, M. Tune the "rainbow" trapped in a multilayered waveguide. *Europhys. Lett.* **2012**, *99*, 57007. [CrossRef]
45. Peng, R.W.; Mazzer, M.; Barnham, K.W.J. Efficiency enhancement of ideal photovoltaic solar cells by photonic excitations in multi-intermediate band structures. *Appl. Phys. Lett.* **2003**, *83*, 770–772. [CrossRef]
46. Matsui, T.; Agrawal, A.; Nahata, A.; Vardeny, Z.V. Transmission resonances through aperiodic arrays of subwavelength apertures. *Nature* **2007**, *446*, 517–521. [CrossRef]
47. Bao, Y.J.; Zhang, B.; Wu, Z.; Si, J.W.; Wang, M.; Peng, R.W.; Lu, X.; Li, Z.F.; Hao, X.P.; Ming, N.B. Surface-plasmon-enhanced transmission through metallic film perforated with fractal-featured aperture array. *Appl. Phys. Lett.* **2007**, *90*, 251914. [CrossRef]
48. Bao, Y.J.; Li, H.M.; Chen, X.C.; Peng, R.W.; Wang, M.; Lu, X.; Shao, J.; Ming, N.B. Tailoring the resonances of surface plasmas on fractal-featured metal film by adjusting aperture configuration. *Appl. Phys. Lett.* **2008**, *92*, 151902. [CrossRef]
49. Zhu, L.H.; Shao, M.R.; Peng, R.W.; Fan, R.H.; Huang, X.R.; Wang, M. Broadband absorption and efficiency enhancement of an ultra-thin silicon solar cell with a plasmonic fractal. *Opt. Express* **2013**, *21*, 313–323. [CrossRef]

© 2019 by the authors. Licensee MDPI, Basel, Switzerland. This article is an open access article distributed under the terms and conditions of the Creative Commons Attribution (CC BY) license (http://creativecommons.org/licenses/by/4.0/).

MDPI
St. Alban-Anlage 66
4052 Basel
Switzerland
Tel. +41 61 683 77 34
Fax +41 61 302 89 18
www.mdpi.com

Symmetry Editorial Office
E-mail: symmetry@mdpi.com
www.mdpi.com/journal/symmetry

www.ingramcontent.com/pod-product-compliance
Lightning Source LLC
LaVergne TN
LVHW070045120526
838202LV00101B/432